I0280148

Glenn Brown

**Healthy Foundations for Houses**

Glenn Brown

**Healthy Foundations for Houses**

ISBN/EAN: 9783337059446

Printed in Europe, USA, Canada, Australia, Japan

Cover: Foto ©Andreas Hilbeck / pixelio.de

More available books at **www.hansebooks.com**

# HEALTHY FOUNDATIONS

FOR

# HOUSES

WITH 51 ILLUSTRATIONS

BY

GLENN BROWN,

*Architect, Associate American Institute of Architects.*

---

" He who builds a fair house on an ill
seat committeth himself to prison."—BACON.

---

(REPRINTED FROM THE "SANITARY ENGINEER.")

NEW YORK:
D. VAN NOSTRAND, PUBLISHER,
23 MURRAY AND 27 WARREN STREETS.
1885.

# PREFACE.

THE following small work is a reprint of a serial paper published in the *Sanitary Engineer* during the year 1884.

All the illustrations are engraved from drawings made for these articles by the author.

New matter and illustrations have been added where it was thought the value of the book would be increased thereby.

A table of contents, with an index to the illustrations has been written, and the text revised and corrected.

Technical terms and phrases have been avoided, where possible, so the essay may be profitably read by any intelligent householder.

I hope that architects and others expert in sanitation may find something new, interesting or of value in its pages.

GLENN BROWN.

WASHINGTON, D. C.,
607 La. Ave.,
1885.

# CONTENTS.

*(The Figures refer to Paragraphs.)*

## I.

### Natural Foundations.



## II.

### Drainage.

## III.

### Foundation Walls.

## IV.

### Miscellaneous Matter.

# INDEX TO ILLUSTRATIONS.

*(The Figures refer to the Number of the Cut.)*

# HEALTHY FOUNDATIONS FOR HOUSES.

## I.

1. *Definition.*—The term or word foundation is used without distinction to mean either the surface on which the building rests or the footing-courses and cellar-walls on which the superstructure is built.

To obtain healthy foundations it is necessary to select or so treat the building site that it will be free or protected from vegetable and animal matter in the subsoil, which in the process of decay would generate poisonous gases; it must also be free, naturally or artificially, from dampness—*i.e.*, there must be no springs, water-courses, or standing water within the area of ground covered by the building.

It being an accepted fact that damp and

polluted soils are unhealthy when used as foundations, it only comes within the province of a treatise of this kind to show how buildings can be protected from the evils which would be produced.

2. *Soil-Water and Moisture.*—Soil-moisture arises where water and air both occupy the interstices of the ground; when the interstices are filled with water to the exclusion of air it is called soil-water. A soil that is full of water acts as a collector and retainer of filth that may soak into the ground, prevents the circulation of air, and thus retards the oxidization or purification of the filthy matter which it contains. Water or moisture in the soil decreases the temperature.

3 *Ground-Air.*—The movements, character, and effects of ground-air have been investigated only to a limited extent, but its important bearing has been clearly shown by the experiments of Prof. Renk.* By his experiments he

* Dr. Renk recommends underground porous flues connecting with the kitchen chimney for taking off the ground air from beneath the building. Such an

finds that the draught or current is from the ground into the house for the greater part of the year, and that it brings with it particles of whatever injurious matter the soil may contain.

These particles would be disseminated through the house by the warm currents of the heating apparatus.

The ground-air is kept in continual movement by the state of the winds and of the atmosphere.

The effect which the wind has upon the ground-air is clearly shown by the following experiment of Pettenkofer.

4. *Pettenkofer's Experiment.* — He takes a long glass tube closed at the bottom and fills it with gravel. Into the axis of this tube he inserts another smaller tube, open at both ends and reaching nearly to the bottom of the large tube. The second tube is attached

---

arrangement might be occasionally effective, but often it would be useless, *i. e*, when the kitchen fire was not burning, when some fire in another part of the house needed the air more than the kitchen fire did, when contrary winds would force it into the house, &c.—G. B.

by a piece of rubber tubing to a glass U-shaped or trapped tube, which is partially filled with water. (See Fig. 1.) "If

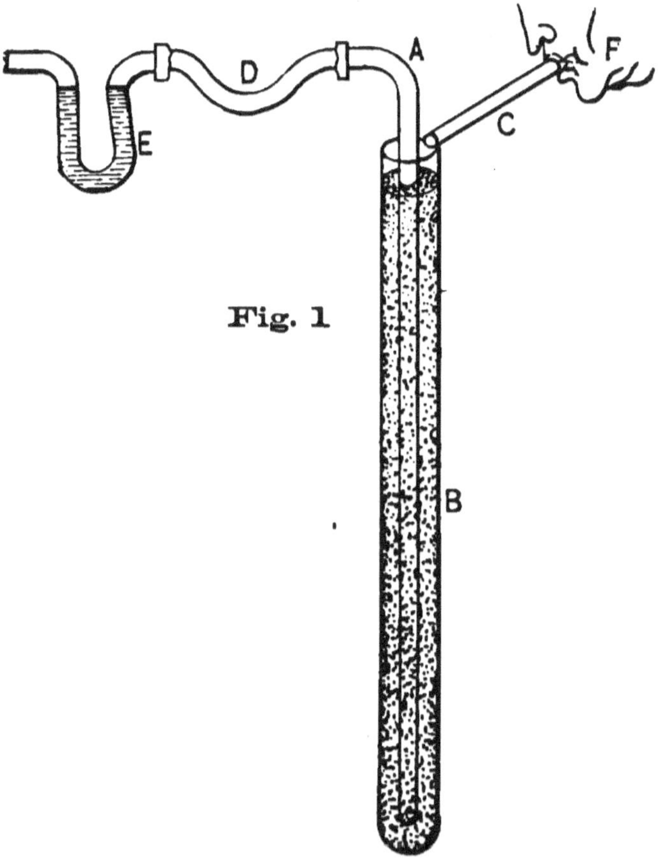


Fig. 1

PETTENKOFER'S EXPERIMENT.

(*a*) small tube, (*b*) large tube filled with gravel, (*c*) tube with person, *f*, blowing through it, (*d*) rubber tube, (*e*) bent or U tube with water in it.

a person blow, as represented in the figure, on the surface of the gravel, the liquid in the U-shaped tube will be seen to alter its position, the level side next to the person who is blowing, becoming lowered, and the other proportionately elevated. The depression of the fluid is caused by the air blown through the gravel, because it ascends from the bottom of the gravel through the small tube, passes through the india rubber tube, and thus reaches the water."

5. If a candle is placed at the mouth of the U-shaped tube, the water having been removed, the flame of the candle will be pressed aside by the current of air. From the experiment it can be easily understood how the ground-air is continually moving by the action of the winds on the surface of the ground, or by the state of the atmosphere.

6. *Geological Formations*.—Natural drainage, and the necessity for artificial drainage of the building-site, depend upon the character and formation of the different strata in the soil and subsoil,

upon their inclination or dip, upon the permeability, porosity, or depth of the substances forming the crust of the earth.

Much of the following description of the geological formations is a rewritten abbreviation from Dr. Parkes' excellent discussion of the subject in his "Manual of Hygiene."

7. *Granite and Trap-Rock.*—In granite, metamorphic and trap-rock formations the slope or natural inclination is great; the water runs off rapidly and leaves the air comparatively dry.

8. *Slates.*—Slates are similar to granite, in their natural inclination being great and also in their being impermeable. In such rocky formations swollen underground brooks or water-veins are often encountered, just after or during rains, which are dry at other times. Architects may be easily deceived by the intermittent character of these streams.

9. *Limestone.*—Strata composed of limestone usually have sufficient slope to carry off water rapidly. In formations

of this character marshes are common and may be found at great heights. The water is retained in large cavities, which are hollowed out of the limestone in the course of ages by the carbonic acid in the rain. These cavities form reservoirs, which should not be overlooked when building upon a foundation of this character. Oölite is the best, and magnesian the worst, form of limestone for a foundation.

10. *Chalk.*—When free from clay and permeable, chalk forms a healthy foundation, although water will rise to a considerable height above the line of complete saturation. Marly chalk is impermeable, damp and cold. When chalk is underlaid by gault clay the building-site will usually be found malarious. Some of the most marshy districts in America are on chalk. Goitre and calculus are not as common as in limestone districts.

11. *Sandstones.*—Strata composed of sandstone are usually permeable and healthy, the ground and air both being dry. The building-site may be damp

when the sand is mixed with clay or when another impervious stratum underlies a shallow sandstone formation.

Sandstone will be wet by capillary attraction one or two feet above the water standing in it. Millstone, grit formations, resemble the granites more than the sandstones in their conditions.

12. *Gravel.*—When of a sufficient depth gravel makes one of the healthiest of all natural foundations, unless, as is sometimes the case, the gravel allows water to rise through it. In case it is surrounded by an impervious stratum the ground-water may have no outlet.

Dr. Parkes in his "Manual of Hygiene" says: "Gravel hillocks are the healthiest of all sites."

13. *Sand.*—As a foundation sand may either be healthy or unhealthy. Pure sand of considerable depth, not retaining moisture, and in which water rises only a short distance by capillary attraction, is healthy. It is unhealthy when underlaid by clay near the surface, when water rises into and through it, and

when it contains organic matter. Impurities are often retained in sandy soils, and the ground-air, which traverses it freely, carries particles of these impurities with it.

14. *Alluvial.*—The so-called alluvial soils are objectionable, as they keep the air moist, have a large amount of organic matter in them, and marshes are also common in such formations. Alluvial soil consists chemically of the same material as the masses from the disintegration of which it has originated. From its formation it has great permeability for air and water, which act as conductors for particles of decayed matter that may be in it. Soil of this character often has alternations of thin, sandy strata and impermeable sandy clay. The vegetable matter contained in it makes the ground air and water impure.

15. *Clay.*—Water neither runs off from or through clay soils, so the ground-air is moist and marshes are common. Clay does not give off vapor as copiously as free open soils, which are

kept wet by the water that underlies them, although evaporation will continue to take place, and the amount of vapor given off will be much greater in the end than from free soils under the same conditions. Organic matter existing in clays undergoes decomposition slowly, but there is no diminution in the amount of carbonic acid gas evolved. Clay is cold, insalubrious, and objectionable as a building-site.

16. *Cultivated.*—In themselves cultivated soils are not unhealthy, nor has manure, when serving its legitimate purpose, been proven hurtful. Such ground has a bearing upon buildings only as it affects the neighborhoods, as it must always be removed from the actual building-site. Rice fields or irrigated land are hurtful, malaria being common in their neighborhood.

17. *Made Ground.*—In cities it is common to find what is called made ground, *i. e.*, lots which are below the established grade and are filled in. With clean gravel or sand as a filling

such ground will be healthy; but when, as is usually the case, they are filled with rags, sticks, leaves, boots, shoes, dead dogs, cats, rats, chickens, etc., they become dangerous to health.

18. *The Soils Filtering Capacity.**—The question of how far the soil and the different ingredients forming the earth filter or remove impurities from the ground water and air, is an important one. How far deleterious matter will travel before it is eliminated from the water and air is a most important question.

19. *Pumpelly's Experiments.* — The National Board of Health, before it was rendered useless by the legislation of a thoughtless Congress, investigated this subject among other important topics connected with the health of the people. This investigation was carried on by Raphael Pumpelly, assisted by George A. Smith. The report of this investigation was published in the thirteenth supple-

* Portions of a paper by the author published in the *California Architect.*

ment of the National Board of Health Bulletin.* In the following lines I will give a short review of the subject. An illustration of one of these experiments is shown in Fig. 2.

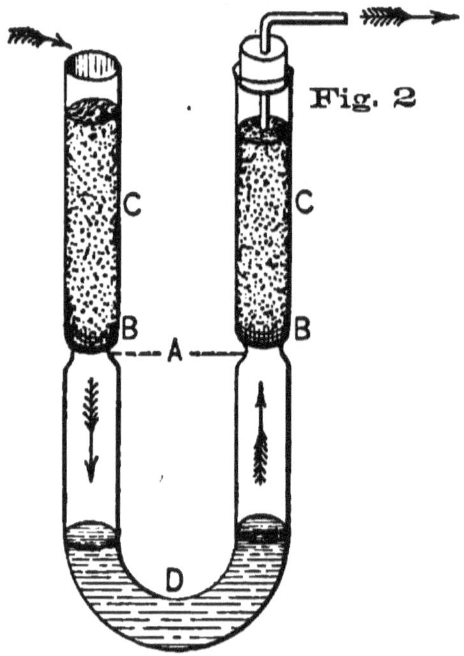


Fig. 2

AIR FILTER.

(*a*) narrow part of tube, (*b*) wire saucers, (*c, e*) sand through which air passes, (*d*) beef infusion; arrows show direction of air.

20. *Air Filters.*—This filter consisted

* Thirteenth Supplement, National Board of Health.

of a column ten centimeters (3.9 in.) high of sand resting on a copper gauze saucer. The sand was 30°, 50°, and 100° fine, *i.e.*, screened through meshes of the corresponding number per inch. In the bottom of this bent tube was some beef infusion, or some other easily putrescible fluid. Air was made to pass through this tube slowly by aspiration. All the filters tried in this way stood the test, the infusion remaining intact.

21. *Water Filters.*—The next experiment shows the filtering effect of the soil on polluted water. The columns of sand used in this case were twenty-two and one hundred feet long, of mixed grain 18° to 100° fine. (See Fig. 3.)

22. "To obtain sterilized columns of such lengths the sand was intensely heated and poured into lead pipe which was then coiled. (Fig. 3.) It was then placed in a furnace and heated to within 250° and 300° C., (482° to 572° F.) and then the lower end attached to a flask containing the infusion." The flask was ventilated through a tube protected

by a sterilized asbestos filter. The whole was then allowed to rest for several weeks, to be sure that the infusion had been properly sterilized, the pipe was then very

Fig. 3

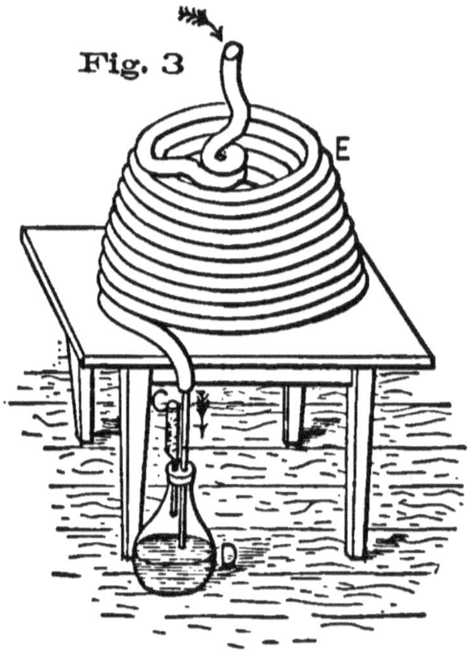


WATER FILTER.

(*e*) coil of lead pipe filled with sand, (*d*) flask with beef infusion, (*c*) ventilated with asbestos filter.

slowly filled with tap water. The first water that passed through both the twenty-two and one hundred feet pipe or columns carried infection with it..

23. *Conclusion from Experiments.*—The inference drawn is that soil is an excellent filter for impure or infected air that may pass through it, but a very poor filter for infected water that may percolate through the soil. The experimenters say; "From these results it appears very clearly that sand interposes absolutely no barrier between wells and the bacterial infection from cess-pools, cemeteries, etc., lying even at great distances in the lower wet stratum of sand. And it appears probable that a dry gravel, or possibly a dry, very coarse sand interposes no barrier to the free entrance into houses built upon them, of these organisms which swarm in the ground-air around leaching cess-pools, leaky drains (sewers), etc., or in the filthy made ground of cities. . . . If the drift of leaching be towards the cellar, very wet seasons may extend the polluted moisture to the cellar walls, whence after evaporation the germs

will pass into the atmospheric circulation of the house."

24. Another series of experiments were made to ascertain the effect of air which either remained stationary or passed in a current over the surface of polluted water. In other words, will ground-air take infectious germs from the surface of polluted water and carry these germs into the house?

25. *Infected Bubbles.*—Fig. 4 shows an apparatus by which filtered air (asbestos filter) is made to pass, by aspiration, along the surface of the infected fluid, and then through the pure infusion at the rate of 3½ liters (7.385 pints) per day. In three flasks, after six months, the infusion was found intact.

26. To try the effect of bursting bubbles on the surface of the water, transferring germs in this manner to the air, other experiments were made. In these experiments "although the formation of bubbles, gentler than in much natural fermentation, the infusion in all three flasks became infected in a few days."

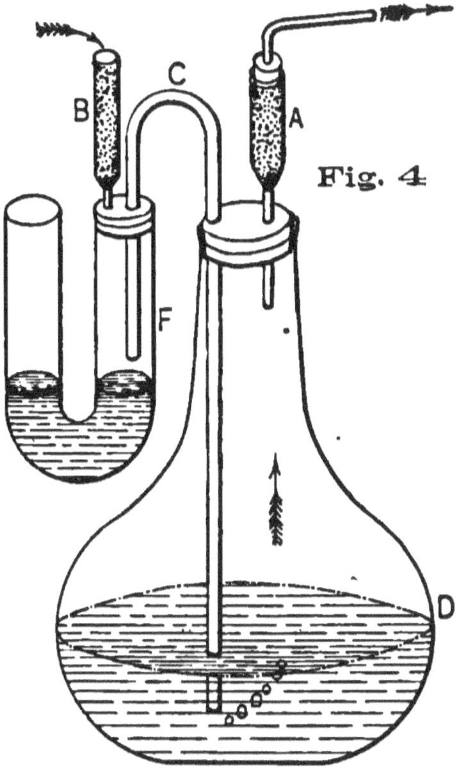


Fig. 4

AIR PASSING OVER POLLUTED WATER.

(*a* and *b*) sterilized filters, (*c*) tube connecting infected fluid chamber with sterilized infusion, (*d*) beef infusion, (*f*) infected fluid: arrows show direction of current.

27. Prof. Pumpelly says in conclusion: "At a normal summer temperature *no germs* were given off from the decomposing liquids *whenever their surfaces re-*

*mained unbroken*, even though in some of the experiments the air was continuously conducted over them in a slow current. When the surfaces of the liquid were broken, however, by the bursting of bubbles, germs were invariably given off and sterilized infusions infected, no matter how slowly the aspiration was conducted."

27*a*. *Disinfecting Microbe.*—Some recent discoveries, or supposed discoveries, in reference to the purifying effects of the soil are described in the *American Architect and Building News* for January 10, 1885.

"It was found a year or more ago that sewage filtered through soil would come out clear and pure, but when this same soil was treated with dilute chloroform, the sewage would pass through without change, the oxidizing or purifying character of the soil being suspended. The soil would not regain its power for some days. This was supposed to show small living organisms, whose activity could be suspended by an anæsthetic, and thus prevent the oxidation of the sewage.

"This theory has now been confirmed by additional observations, and the little creature which converts into a fixed and harmless salt the putrifying impurities of such sewage as it can reach, is believed to be a micrococus somewhat resembling the yeast plant.

"Many and varied tests have been made to determine the conditions under which the disinfecting microbe lives and acts, and a good deal has been learned about its habits. It is found that it flourishes best and is most efficient at a temperature of about ninety-eight degrees Fahrenheit. . . . At higher and lower temperatures it becomes more feeble and ceases altogether near the freezing point, or above one hundred and thirty degrees. Experiments to show its distribution in a clay soil, show that it is most abundant in the upper six inches, but is found to a depth of a foot and a half, below that depth it cannot live. Acid solutions are not acted on by them."

28. *Subsoil Water.* — A site which would otherwise be good is often water-

logged by having an impervious stratum around and beneath it, the water standing at the height of the lowest outlet (Fig. 5).

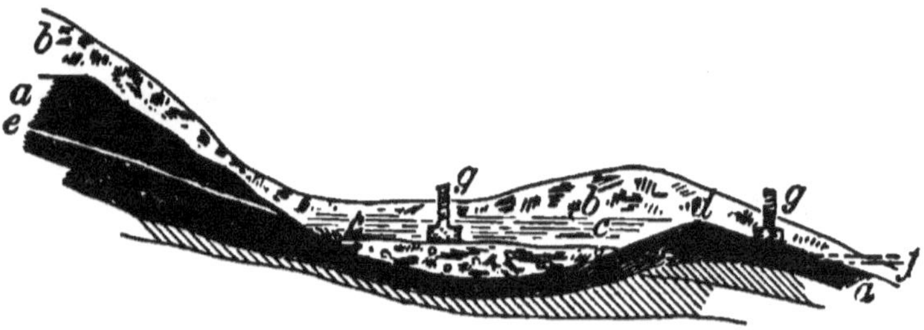


Fig. 5.—Subsoil Water.

*a*, rock subsoil; *b*, common earth; *c*, subsoil lake; *d*, outlet; *e*, water-vein or inlet; *f*, dotted line shows artificial outlet; *g*, positions which foundation walls may assume.

Below, as on the surface, water is to be encountered in the form of streams, springs, and lakes. The height of subsoil water varies according to circumstances, depending on the rain-fall, the height of streams in the neighborhood, and in some instances on the rise and fall of the tides.

29. These streams or springs may be dry at some seasons of the year, so it behooves the architect to examine and see

in what direction impervious substrata may incline, and at what depth they exist below the ground. If a house is to be built on the side of a hill, and stone is seen cropping out above and below the building (Fig. 6), we may conclude that if any por-

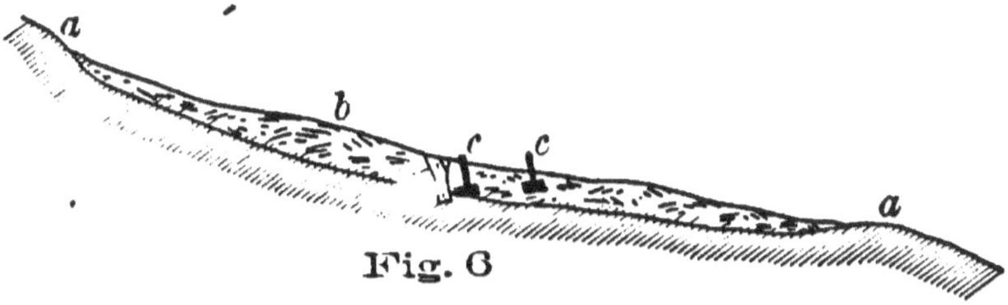


Fig. 6

Sloping rock-stratum cropping out at *a a*; *b*, common soil; *c*, foundation-walls.

tion of our foundation-wall or cellar bottom comes within two feet of this rock or other impervious substratum, then the inmates would be troubled with damp walls and cellars, unless the proper precautions have been taken.

30. The natural foundation in the case of peat-sand or soft clays, should be drained before the foundation-walls are commenced, when drainage is necessary, because in lowering the water-level boggy ground and soft clays are made less bulky,

and the flow of water causes more or less movement in loose sandy soil that might easily become dangerous to the superstructure.

## II.

### DRAINAGE.

31. It is necessary to keep the subsoil water, whenever it stands on or rises into our building-site or cellar, at a depth below the foundation or bottom of the cellar sufficient to prevent the water rising by capillary attraction into the wall or floor.

32. *Drains.*—Artificial conduits for the purpose of carrying water from the subsoil are called drains. The word drain has a meaning distinct from that which should more properly be called a sewer, the first being a duct to draw off pure water, and the latter a conduit for waste matter or sewage.*

---

* The word *drains* and the French word *trainer* have the same meaning; but Mr. E. F. B. Denton is apparently wrong when he states in his book that the English word is derived from, or is a corruption of, the

33. In country houses it may be necessary or beneficial to drain the ground for some distance around the building, while in the city it will be necessary usually to confine the drainage to the building-site proper.

For these drains an outlet must be found. In the country it is easy to find a low point or water course; in the city the drain must be properly connected with the sewer. The manner of making, laying, and jointing drains for the purpose of health are similar to the methods adopted for agricultural works.

34. *Ancient Roman Drains.*—The underground conduits discovered by De Tucci, the archæologist, in the Roman hills, and called *cuniculi*, were examined by Tomassi Crudelli and found to be ex-

---

French. According to W. W. Skeat, in his recent work on etymology, drain is the old Anglo-Saxon word drehnian, or drenian, in which the first syllable, dreh-drah-drag, to draw or pull forcibly, indicates the service a drain renders in extracting the water from the ground. Sewer is derived from *essuier* of the old French (mod. French *essuyer*). The old English form was *essewe*, the first syllable, *es*, having been dropped in later years.—G. B.

clusively for the purpose of draining the hills. In writing of the open walk (Hyphræthum) or promenade situated at back of theaters, Vitruvius describes the Roman manner of making drains: "In order to render the walks firm and free from damps, the earth must first be dug out to a proper depth, and drains of brick-work constructed on each side, and *channels left in the walls*, with a gentle inclination from the walks into the drain. The whole space may then be filled in with charred wood, and the surface covered with sand, which, when leveled, forms the walks. Whatever water falls will either be carried by the channels into the drains or will filtrate through the charcoal leaving the surface of the walls firm and free from moisture."

35. For a historical review of drainage in this country and in England, I will refer the reader to H. B. French's work on the subject.

36. *Construction of Drains.*—They may be formed of earth, wood, stone, or tile (terra-cotta), the latter being decidedly

the best; but it may be expedient sometimes, for economy, from necessity, or as a temporary arrangement, to use the drains first mentioned.

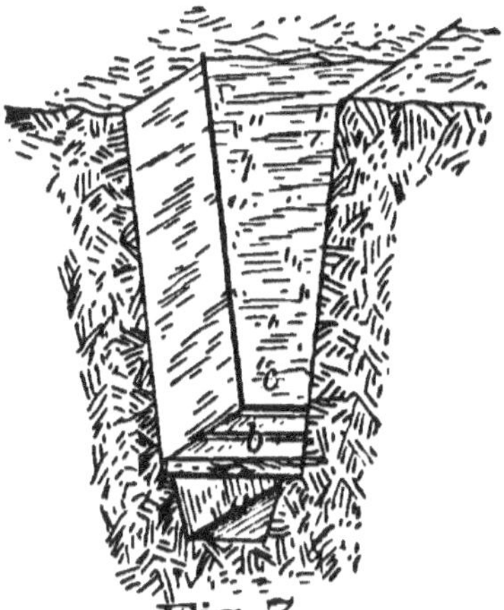

Fig. 7

EARTH-DRAIN.

*a*, drain; *b*, boards; *c*, clay covering.

37. *Ditch or Earth Drains*—Such drains are necessarily open, unless covered by some unyielding substances, such as short pieces of boards or pieces of stone that may be picked up on the premises. The part that is to serve as the drain is smaller than the ditch above, having two shoulders left to receive the

boards, planks, or stones which are to cover the drain. The earth is then thrown in on top, clay being first packed next to the boards or stone, so sand and loam will not get in and fill up the drain (Fig. 7).

38. *Wooden Drains.*—When the ditch has been excavated to the depth desired, small poles are laid in the bottom and parallel to the length of the ditch. If the wood is first charred, as in the case of the Roman drains described above, it will last a long time (Fig. 8). Box-drains

FIG. 8.—POLE-DRAIN.

*a*, pole; *b*, clay cover.

may be formed by nailing short pieces of board, the width of the intended drain, across horizontal pieces which run parallel with the ditch (Fig. 9).

FIG. 9.—BOX-DRAIN.

*a*, box; *b*, clay cover.

The three drains described above may be temporarily used, or used where the better stone or the best tile drains cannot be obtained.

39. *Stone Drains.*—The ditch is partially filled with loose stone (Fig. 10). Large flat stone are sometimes placed so as to form an open conduit at the bottom

Fig. 10

STONE DRAIN.

(Fig. 11). When they are to act as inter-

FIG. 11.—STONE DRAIN.

cepting-drains—that is, catching and carrying off water that may be flowing along impervious strata toward the building—the stone would be carried within a short distance of the surface. In this manner all water flowing toward the building, either from the surface or along a rocky or clayey formation, would be intercepted and trickle down into, and be carried off through, the drain. When the drain is to carry off only subsoil water it must be covered with compact puddled clay, as it is desirable to prevent water entering at the top and carrying with it sand and loam, which would in a short time fill up the interstices between the stones. The water should come in from below. Stone drains are well adapted for intercepting-drains, and also as drains situated beneath or on one side of the foundation-walls.

The outlet, or manner of connecting with the sewer, would be the same as in the case of a tile drain.

Bricks are not suitable for the purpose of making drains.

40. *Tile Drains* are short pieces of

hollow, porous terra-cotta, or burnt clay, and are far superior (and to be used except in special cases) to the drains already described. They carry off the water with greater ease, rarely, if ever, get choked up, and require only a slight inclination to keep the water running through them.

Drain-tile may be made from the same clay that will make good brick. This clay contains in a hundred parts:

Silica, from 55 to 75 parts;
Alumina, from 25 to 35 parts;
Magnesia, from 1 to 5 parts,

mixed or combined with variable, or 0, quantities of lime, potash, and oxide of iron. The clays are not often adhesive when taken from the ground, and they have to be tempered, or thoroughly mixed, it being sometimes necessary to add foreign substances to make them useful. The pipes made are porous; but it is not intended to have the water enter through its shell, but through the joints.

41. The Romans are known to have used unglazed terra-cotta pipe; but these pipe are supposed to have been intended

for the purpose of sewerage, and not for the purpose of drainage.

42. Drain-tiles are made in a variety of patterns, as horseshoe, sole, double-sole, and round-tile, the name being derived from the shape or appearance of their cross-section. Drains have been formed on the bottom and in one piece with the stone or earthenware sewer-pipe.

43. *Horseshoe Tile*, which were originally open at the bottom, were, or are, placed with the open end down on the bottom of the ditch. They were so far superior to stone drains that they had an extended use; but they are much inferior to the round tile (*a*, Fig. 12). The horseshoe tile is also made with a bottom to it (*b*, Fig. 12).

44. *Sole-Tile* have a flat bottom or foot piece formed on the tile for the purpose of giving them an even and flat surface to rest upon. The form of the tile defeats the end for which it was designed, as the clay in baking is distorted by the unequal thickness, which causes the top to dry and contract much sooner than

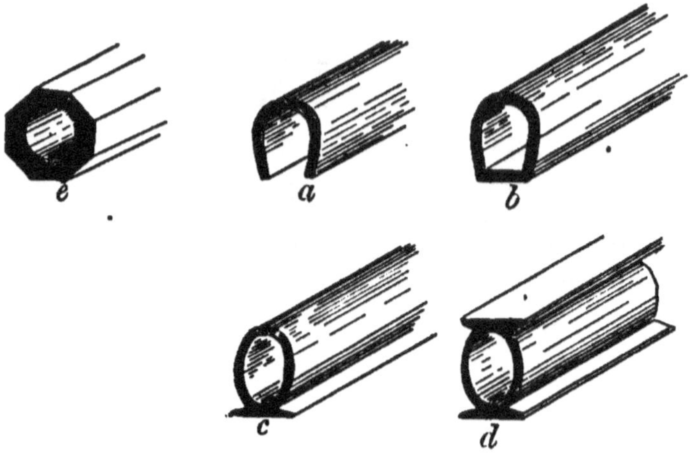


Fig. 12.

*a*, horseshoe tile; *b*, horseshoe tile with bottom; *c*, sole tile; *d*, double-sole tile; *e*, round tile, octagon-shaped outside.

the bottom. With the tile misshapen it is impossible to give them proper joints (*c*, Fig. 12).

45. *Double-sole Tile* have the flat footing formed on both sides. They are not often distorted, and may be laid either side up; but are heavy, which adds materially to the cost of transportation. The joints are made more easily than in the case of sole-tile (*d*, Fig. 12).

46. *Sewer and Drain Tile combined.*—Sewer terra-cotta pipe has been manufactured with two compartments, one for sewage, and the other or bottom compartment for pure water from the ground. Fig. 13 is illustrations of pipe for this

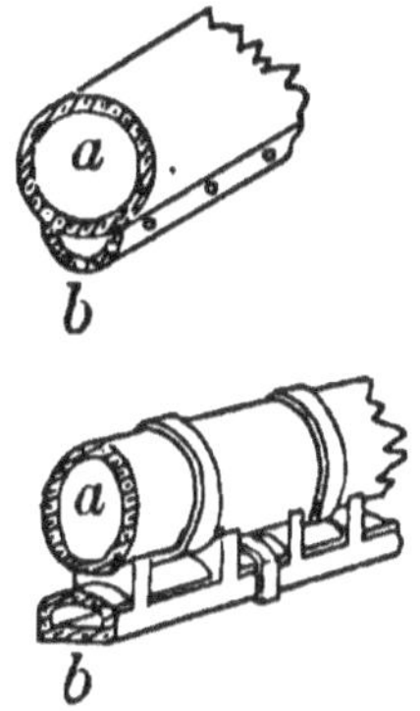


FIG. 13.

*a*, sewer; *b*, drain.

purpose, the drain-pipe having small openings through which the water can enter. In practice a drain-pipe of this kind would be certain to collect sewage-water from the imperfect joints in the upper compartment.

Although combination-tiles are sometimes used to carry off ground-water while the sewer is being laid, it will usually be better to drain the ground separately.

47. *Round Tiles*, as the tiles are called that are circular in cross-section, with collars over the joint of the small tile and without collars for the large tile, form the best conduits for subsoil drainage.

The inside diameters of these tiles usually vary from $1\frac{1}{4}''$, $1\frac{1}{2}''$, $2''$, $2\frac{1}{2}''$, $3''$, to 6 inches, but they are manufactured as large as 8 inches. Pieces of the larger pipe serve as collars for the smaller sizes. They are in length 12, 14, and 24 inches, and in thickness of shell from a quarter of an inch to one inch.

49. The collar which encircles the joint of the small tile allows a large opening, and at the same time prevents sand and silt from entering the drain.

The large tiles do not need the collar at the joint, as the larger diameters are only needed (or used) when there is an abundant flow of water—a flow sufficient to carry off any silt that may get into the pipes. Large drains sometimes have a bell similar to sewer-pipe. In Fig. 14 is

FIG. 14.

*a*, small drains; *b*, collar; *c*, main drain; *d*, collar for joining branch; *e*, branch fitted into collar; *f*, Y branch; *g*, clay packing.

shown a tile drain in position, the small branches having collars and the main drain being without them.

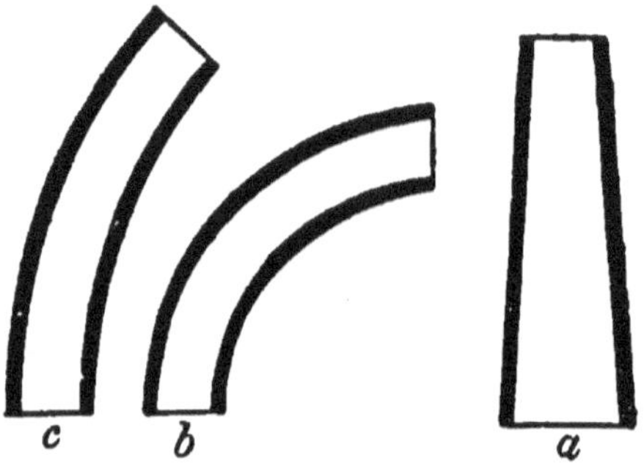


FIG. 15.—ROUND TILES.

*a*, reducer; *b*, quarter bend; *c*, eighth bend.

50. *Jointing Tiles.*—Not many years ago, when it was necessary to change the course of a line of drains or to connect a branch into another line, it was necessary to chip with a small pick-hatchet until they joined at the angle or fitted into the hole. The improvement on this method consists in having eighth and quarter bends when the course of the drain is altered, and Y and T branches to connect the lateral with the main drain.

51. Where the Y branches are not easily obtained the lateral is joined to the main drain by chipping the lateral, so

that it can fit into a hole shaped in a long collar to receive it (*d* Fig. 14).

The pipes which fit into the collar are only to be brought to the edge of the opening. Connections of this kind must be made on the top of the drain (Fig. 16),

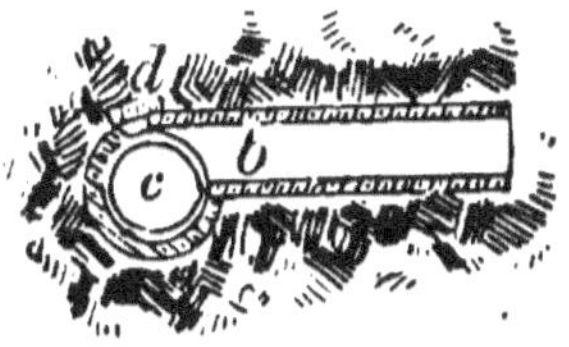


FIG. 16.—CONNECTION.

*a*, collar; *b*, lateral; *c*, main; *d*, cover to point.

otherwise the irregular joints would form points of deposit for silt and sand, while Y branches bring the current from the lateral drain in at the bottom. In this manner the drainage system is more thoroughly scoured, and there is not as great a necessity for the use of silt-basins.

When the joint is formed by the pick it is necessary to cover it with pieces of tile, tin, leather, or paper, and have compact clay packed around it. Reducers or increasers are used where the diameters of pipes that are in a straight line are to be changed.

52. *Large Tiles.*—The tiles with large diameters are simply placed end to end, and to protect them from and prevent the earth from running in at the joint, the ends must come within at least a quarter of an inch of each other at the top, being beveled off with pick when it is necessary. All tiles in manufacturing dry more quickly at the top than at the bottom, and so are liable to be shorter on one side than on the other. This is particularly the case with large tiles. When laid in the ditch they may be held in their place, when it is necessary, by small stones. The joints, to prevent loose earth falling in when they are without a collar, may be covered by tin bent around the pipe, or scraps of leather. Grass, rope or paper may sometimes be used. With the former, it must not be too thick; with the latter there must only be a small amount. Stiff cotton cloth is sometimes wrapped around the end of drain-pipes in the place of collars. Sub-

stances that decay are not as desirable around the joint as the terra-cotta collar, because the mould will eventually find its way into the drainage system.

53. *Planning or Mapping Drainage System.*—Before laying the drains the ground should be examined and the levels

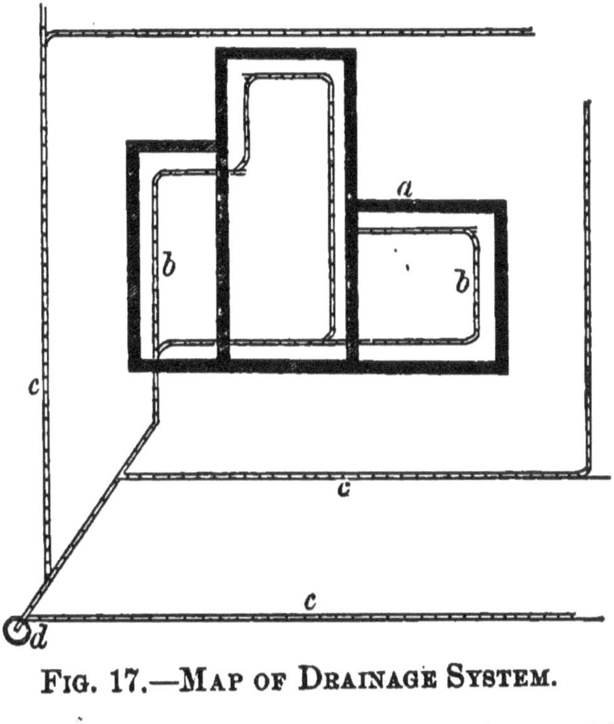


FIG. 17.—MAP OF DRAINAGE SYSTEM.

*b*, drains within the building; *c*, drains outside of the building; *a*, plane of building; *d*, silt-basin connected with drain to outlet.

taken, the geological formation studied, and the outlet or point of sewer connection fixed upon; then a careful map or plan should be drawn (Fig. 17), noting the direction and inclination of drains, curves, connections, silt-basins, and outlets, as well as any peculiarity in the geological formation.

One set of drains of the smallest size is all that will be required on the building-site itself or beneath the foundation-walls, unless the building is very large, the ground of the most retentive character, or the building-site filled with springs.

54. *Height of Ground Water.*—Fig. 18 explains how the height of ground

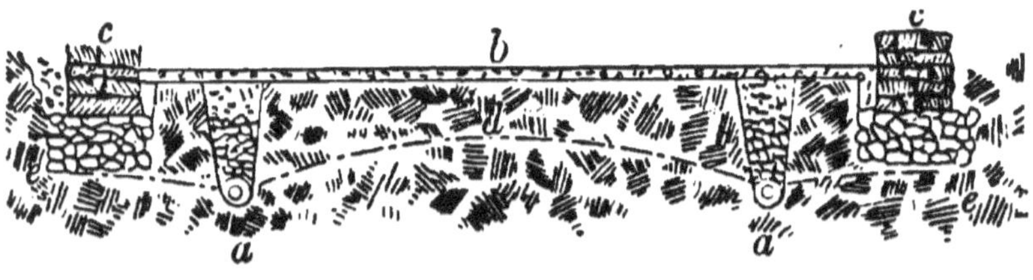


FIG. 18.—GROUND WATER LINE.

*a*, drains; *b*, concrete; *c*, foundation; *e*, stone drains; *d* highest point at which water stands.

water is lowered by drainage, the highest point being midway between the drains and the lowest point where the water enters the drain itself.

Water, except in intercepting-drains, should enter from beneath, when it will be clear and free from obstructions, otherwise the drain will be filled with earth, sand, and silt.

55. *Depth of Tiles.*—Josiah Parkes considers four feet the minimum depth at which drain-tiles should be placed below the surface.

According to Prof. Mapes's rule, tiles put down three feet deep should be laid twenty feet apart, and the distance apart should be doubled for every additional one foot in depth. Tiles laid four feet deep should be forty feet apart; five feet deep, eighty feet apart, and so on.

The deeper the drain the more quickly the water runs into and through it, as it is acted on by the additional weight of water above.

In draining for foundations, of course the depth of the foundation-walls and cellar or basement floor must be taken into consideration.

56. *Tile Laying.*—After laying out and staking the course of the ditches, the intended fall or inclination must be gauged at the upper and lower ends of each straight section of drain-pipe. A board, nailed to two upright pieces which straddle the ditch and are driven into the ground to the proper depth, serves as a basis to measure from at any point on the line when a cord is stretched from one to the other, the cord giving the proper inclination at each point in the drain. The fall required for a line of tiles is much less than would be generally supposed. More than one foot in a hundred it is unnecessary to seek; that much is desirable, and a greater inclination is not objectionable. Six inches is sufficient where the work is carefully executed.

57. Where the fall or grade is less than one foot in a hundred, or in case the foundation is bad, being in contact with quicksand or marshy ground, it is best to open the whole line of ditch, and com-

mence laying at the upper end, continuing it down-stream. When there is a fall of as much as one foot in one hundred feet, or a greater inclination, it will be best to adopt the English method of laying the drains—that is, start at the lower or outlet end and lay them up-stream. By this method the workman never touches the bottom of the ditch, so the ground is in the best shape to receive the tiles.

58. Drainage of the building-site and grounds in the country is in fact agricultural drainage, carried on for the preservation of health; in this case it would be wise to employ regular drainers.

The tools named are English agricultural draining-tools, suitable for alluvial or peat soils, and the method described is the English manner of laying tile in soils which are easily worked.

59. *The tiling implements* used are two ditching spades, one twenty inches long in the blade, six inches wide at the top, and four inches wide at the point; the other a little shorter, four and a half

inches wide at the top and three inches at the point. These spades have steel points and must be kept sharp (*a*, *b*,) *Fig.* 19.

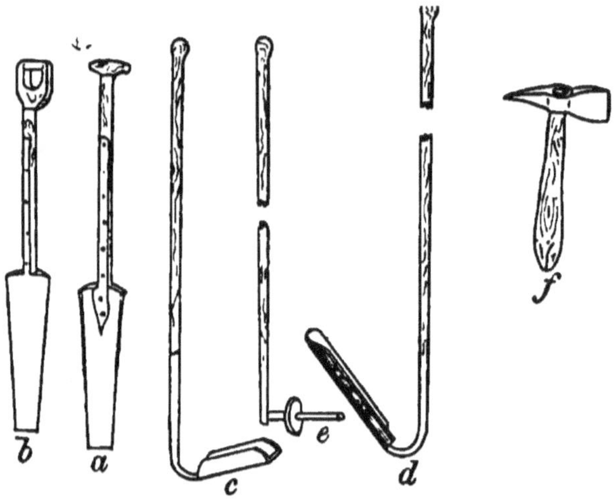


FIG. 19.—TILING TOOLS.

*a*, large ditching-spade; *b*, narrow ditching-spade; *c*, flat scoop; *d*, cylindrical scoop; *e*, tile-layer; *f*, pick-hatchet.

Scoops, one with a flat bottom and rectangular sides, four inches wide, and the handle so arranged that the workman can step backward as he uses it (*c*).

A bill-scoop, with the section of its bottom segmental, three inches wide, and its handle arranged in the same way as that of the flat scoop (*d*), *Fig*. 19.

The "tile-layer" is simply a short rod placed in a handle, and forming a right-angle projection from it, having near the handle a circular fender that will keep tile a proper distance from the handle (*e*), *Fig*. 19.

60. *English Method of Tile-laying.*—When a line has been stretched, the sod removed, and a narrow ditch dug to the depth of eighteen or twenty inches, and fifteen or sixteen feet long, then the wide ditching spade (*a*), *Fig*. 19, is used. It may be necessary to use a common pick in very hard ground.

Any loose earth that remains is scraped up by the flat-bottomed scoop; the workman steps backward, and when the scoop is full swings it out on the side and empties it. In this way a smooth bottom is formed about three feet from the surface, on which the workman can stand. Taking the narrow spade, he commences at the lower end of the ditch and makes it about one foot deeper. The workman always faces the point of commencement and works backward on the smooth bottom previously prepared. When the ditch has been dug to its desired depth the bill-scoop is used to clear it out and give the ditch a cylindrical bottom. The workman is never less than a foot above the bottom, so it is in a much better condition than when the workman walks on the bottom of the ditch and scoops up earth with a spade formed for that purpose.

Tile is laid in place by means of the tile-layer in the hand of a workman standing on the surface astride the ditch. The collar is placed in position on the end of the tile before it is lowered into place. The free end of the tile is inserted into the collar of the tile that is already in position. When the pipe has been laid, then the part of the removed soil that has the most clay in it is thrown down carefully on the tile and tramped into place, leaving the collar of the tile last laid uncovered. The ditch is then filled and tramped or well pounded into place. Another section of fifteen or sixteen feet is then laid in the same manner. There is little danger of dirt or water entering the pipes, as such a small section of the ditch is open at one time; but should there be a large amount of water in the soil, or a probability of rain, the end of the tile should he plugged to prevent the entrance of dirt. This method is graphically described by Col. Geo. E. Waring, Jr., in his work "Drainage for Profit, and Drainage for Health."

61. *Obstructions.*—Vermin, roots, and earth or silt cause obstruction in drains. Rats and field-mice can enter only through some opening, such as an outlet, silt-basin, or sewer-connection, all of which must be properly protected by gratings or screens.

Roots sometimes enter pipes and fill the whole bore with their fibers. According to Gisborne, roots do not, apparently, enter unless there is a perennial stream running through the pipes. When the line of tile passes by a tree, the joints may be filled with Portland cement and sand; asphalt is said to prevent roots entering the drain. As tight joints affect the efficiency of the drains, it may be necessary to sacrifice the tree by cutting it down.

62. *Silt-basins.*—Silt is earth that is carried in water by mechanical suspension, and it would be deposited where imperfect joints in tiles or eddies occurred. Compact clay yields very little silt, quicksand a great deal. The longer the system is in operation the less danger there is of silt.

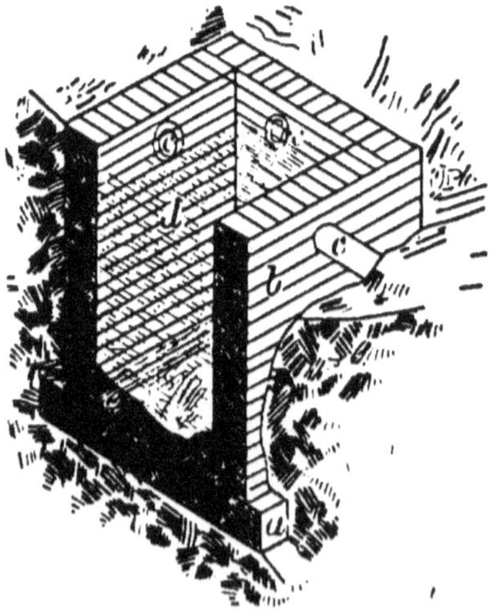


FIG. 20.—BRICK SILT-BASIN.

*a*, stone slab; *b*, brick-work; *c*, drain-tile; *d*, water; *e*, deposits.

Silt-basins are chambers making a break in the line of tile, with their bottoms below the established grade, into which the water would flow and deposit its silt and sand or other particles of heavy foreign matter. A silt-basin may be small, made from a single 6-inch piece of terra-cotta pipe (Fig. 21); or large, made in the form of a brick chamber, and holding many gallons (Fig. 20).

They may break the connection in one line of drains, or they may form the central point for three or four lines (Fig. 20).

63. They form a convenient manner of connecting the drains from the grounds and foundation-drains (the two might necessarily be at different level) with the same outlet.

The pipe or brick-work is set on a flagstone bottom, and carried to within a foot of the surface, and covered with a piece of stone, or carried to the surface and covered with an iron grating.

64. Near the building where the earth has been disturbed, and where intercept-

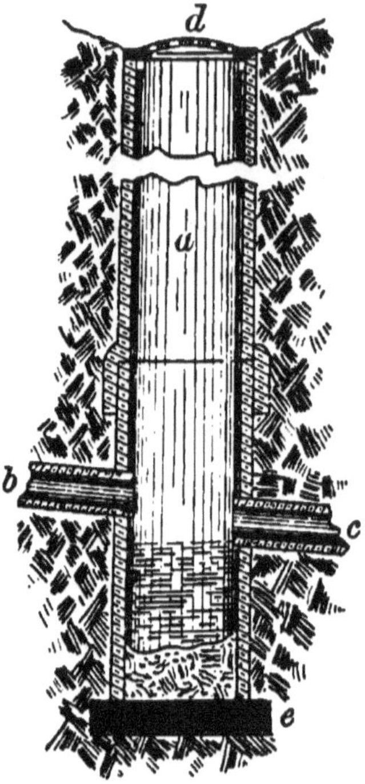


FIG. 21.—PIPE SILT-BASIN.

*a*, terra-cotta pipe; *b*, inlet; *c*, outlet; *d*, iron grating; *e*, stone.

ing drains enter and take the water from near or on the surface, there is liable to be more or less dirt. In this case, and where there is quicksand, silt basins may be found useful, although good authorities think they are not needed in agricultural work.

65. *Springs.*—Where springs occur on or near the building site, then the earth around them must be dug out and the space filled in with loose stone. A tile drain run through it must connect with the drainage, or by a proper trap with the sewerage system (Fig. 22). A section of terra-cotta pipe six or eight inches in diameter may be placed in the opening, being surrounded by broken stone

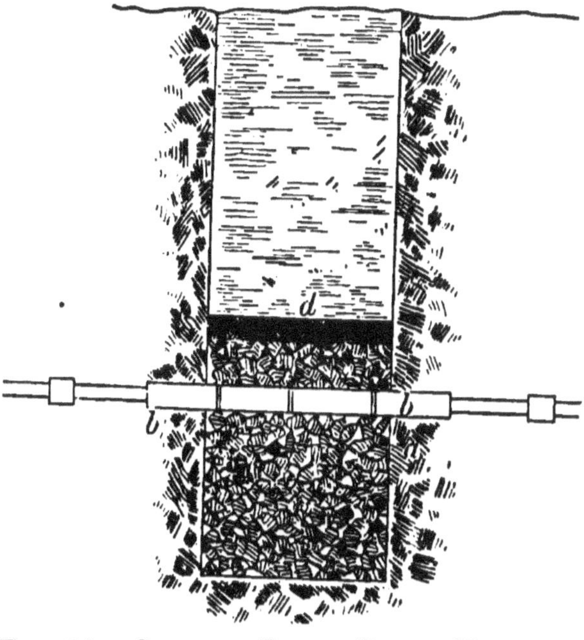


FIG. 22.—SPRING—LOOSE-STONE FILLING.

For reference letters, see Fig. 23.

and covered with a flat stone or tile, and the earth thrown in on top. The terra-cotta pipe acts as a reservoir into which the water can collect (Fig. 23).

FIG. 23.—SPRING—TILE RESERVOIR.

*a*, loose stone; *b*, drains; *c*, terra-cotta pipe; *d*, stone slabs.

66. *Outlets.*—In the country the drains are conducted to a creek or some low point, from whence the water can run off

from the building or into a natural watercourse.

The outlet may be formed by building a dwarf-wall of brick or stone, whichever is cheapest or most convenient in the locality, against the earth embankment, from which the outlet will project. A stone let into the wall, and having an iron grating covering a hole cut in the center of it, serves as an outlet for the drainage system (Fig. 24). This grating

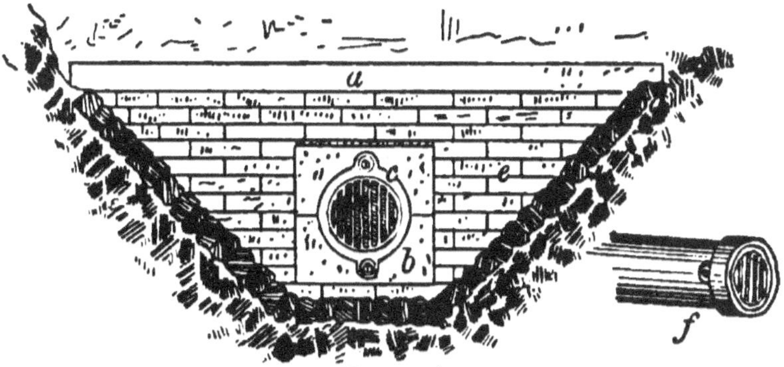


Fig. 24.

*a*, coping; *b*, stone outlet; *c*, iron grating; *d*, paving; *e*, brick wall; *f*, earthenware outlet and grating.

is necessary to prevent vermin entering the drain pipes, building nests, and thus choking up the openings in the tile.

There is an earthenware grating manufactured which fits into an earthenware pipe, that can be used instead of the outlet previously mentioned (Fig. 24—*f*). Where the water runs out on the ground, the surface must be paved with small stones, or a large, flat stone must be used to prevent the water wearing the earth away.

67. *Sewer Connections.*—A very important feature, probably the most important feature, in soil or subsoil drainage, where the drains are beneath the building or are connected with walls in any way, is the manner of connecting the drains with a sewer. It is usually a necessity in the city to have the drain-water find its way into the sewer; but they must never be directly connected, as the sewer-air (so-called sewer-gas), with its organic or disease germs, would be drawn by the warm currents of air directly into the house. The drain and the sewer should be trapped separately. These traps should not be connected, but have an open and venti-

lated chamber between them. I illustrate four methods of making the connection with a sewer, two with mechanical valves, and two with water-seals only between the sewer and the drains.

68. In drains where the flow of water is intermittent, whether it is caused by the seasons or otherwise, the water in the trap would be liable to evaporate and leave the pipe open for the ingress of sewer-air. In a case of this kind it

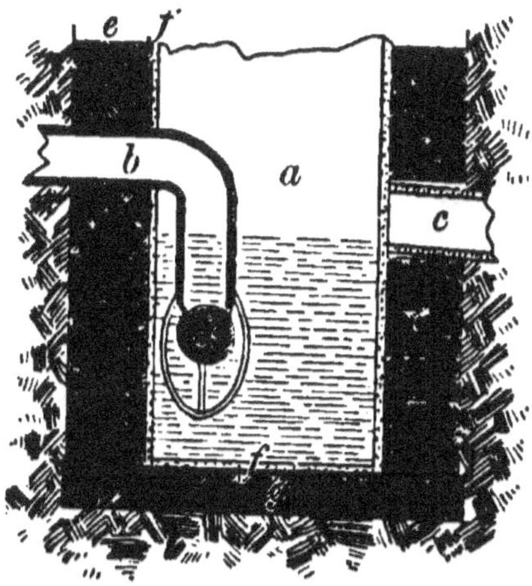


FIG. 25.—VALVE-CONNECTION.

*a*, reservoir; *b*, drain; *c*, sewer; *d*, valve and guard; *e*, brick-work; *f*, cement lining; *g*, flagstone.

seems proper, if at all, to use a mechanical valve in connection with the water-seal. Mechanical valves are never to be implicitly relied on, and the water-seal, unless the flow is continuous, should always be between eighteen inches and two feet (Fig. 26) in depth.

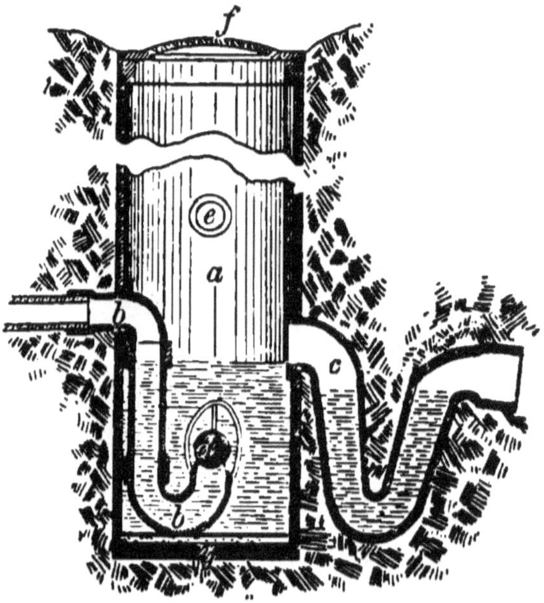


FIG. 26.—VALVE-CONNECTION.

*a*, terra-cotta receptacle; *b*, inlet from drain; *c*, trap and outlet to sewer; *d*, rubber ball with guard; *e*, rain-spout; *f*, grating; *g*, bottom of concrete, cement top.

69. *Valve Connection.*—A brick manhole is sometimes built having its sides and bottom thoroughly plastered with Portland cement and sand, so none of the water will filter through into the soil. The water, unless frequently renewed, will disappear with sufficient rapidity by means of evaporation. The outlet from the manhole into the sewer should have a deep running trap (*c*, Fig. 26), so, in case the trap on the drain should become inoperative, then the running trap would be interposed between the sewer and the house, and it would also prevent sewer-air passing out at the grating into the yard or front parking. It is better to ventilate the sewer further from the house than these manholes are usually situated. Fig. 25 shows an outlet with a ball valve that is held against its seat by the pressure of the water. This ball must be light enough to float. Such valves are only useful as long as the water remains in the receptacle. Fig. 26 represents a sewer connection formed in a large, glazed terra-cotta pipe. The top

is covered with an iron grating, the bottom filled with concrete and cemented, the inlet from the drain and the outlet to the sewer being manufactured of cast iron. The drain outlet runs down the side of large terra-cotta pipe, and forms a return bend at the bottom. The valve seat should be of brass and brought to a knife edge. A hollow brass or copper ball would form a good valve. A joint with valve of this kind would usually remain perfect, even after the water had evaporated. It is a good plan to run the rain-spout from the roof into these receptacles, as they help to keep the water-seals perfect. A trap of this kind without the return bend and ball valve would form an excellent double water-seal trap between the drain and sewer.

70. *Water-seal Connection.*—Fig. 27 represents a double water-seal trap, formed by sinking ten or twelve inch terra-cotta pipes with cement bottoms, and cutting the entrance for the drain into one and the exit for the sewer into the other. They must have an opening

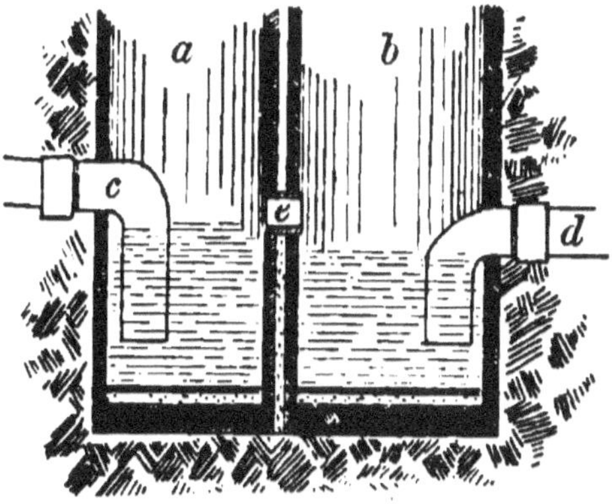


FIG. 27.—WATER-SEAL CONNECTION.

*a*, large pipe-trap for drain; *b*, trap for sewer; *c*, drain; *d*, sewer; *e*, connection between traps.

between them above the outlet to the sewer. The pipes, both from the drain and the sewer, turn at right angles and run down into the water to within four or five inches of the bottom of the terra-cotta pipe. In this way both connections are trapped and have an opening above for ventilation. In case the traps should become unsealed, a current of air could flow from the sewer to the outer air, or from the outside air into the drain.

71. *Hellyer's "Triple Ventilating Trap,"* being manufactured in one piece of earth-

enware, forms a good disconnecting trap between the drainage system and the sewer when the flow of water is continuous. The seal being shallow and surface large, it would soon be broken by evaporation. (Fig. 28.)

A manhole built of brick, or a large piece of terra-cotta pipe, may be placed

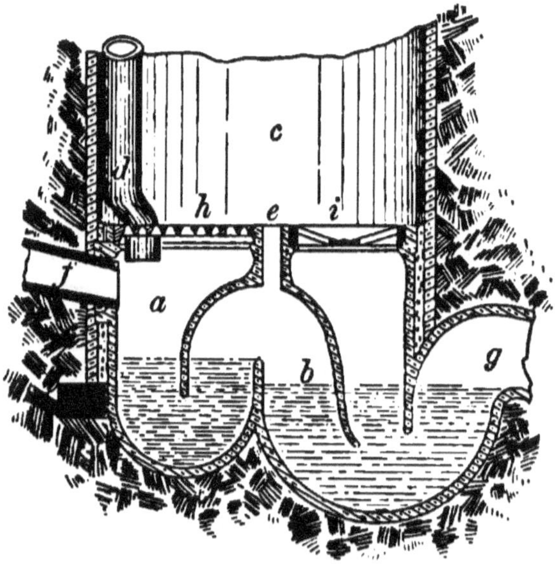


Fig. 28.—Double Trap in One Piece.

*a*, trap for drain; *b*, trap for sewer; *c*, manhole; *d*, rain-spout; *e*, vent between traps; *f*, drain; *g*, sewer; *h*, open grating for drain-trap; *i*, open grating for sewer-trap.

around this trap, and so connect it with the surface.

The construction of this trap is fully explained by the illustration (Fig. 28). It is advisable, where the rainfall and the flow of the subsoil water are uncertain, to have a driblet from the water supply, or a waste pipe from which only clean water would run, to empty into the trap and keep the water seal intact.

## III.

### FOUNDATION-WALLS.

72. *Selection of Site.*—Whether the building shall be placed on the top of the hill, on the side or slope, or in a valley, is often more a matter of necessity than of choice. The side of the hill is usually to be preferred. Hill tops are liable to be of a rocky formation, and to have sheets of water held in place beneath the soil, while valleys are generally the seat of water-courses.

When in doubt as to the character of the ground, it is best to sink wells, or make borings in different parts of the building site, from which the architect will be able to judge the character of the substrata, as well as the way in which they incline.

73. There is generally a choice possible in the location of the building, a few inches difference in depth, or a few feet from a ledge, may constitute the difference between a wet and a dry cellar. In some cases the house may be with advantage placed above the ground and terraced. When it is impossible to remedy the evil by a choice in position, then it becomes necessary to drain, or take some of the precautions which are to be described, or to combine drainage with some one of these methods. Water, ground-air, and with it probably organic germs, may enter the building through the foundation-wall from the side, through the footing courses by capillary attraction, or directly through the cellar floor from beneath.

74. *Absorption of Moisture.*—Brick and stone work, according to their char-

acter and quality, absorb more or less water. The greater the air-spaces which are between the impervious grains of all building brick or stone, the more water can be held in suspension. According to data given by Mr. Eassie, a cubic foot of common brickwork will absorb about ten pints, sandstone about eight pints, and granite a little less than three pints of water.

Edward Cresy, in the Encyclopedia of Engineering, gives an extended account of the capacity of different English stones for absorbing water.

75. *Gilmore's Experiments.* — From the interesting experiments of Col. Q. A. Gilmore, on the absorption of moisture by building-stones in the United States, I take the following information.

Granite absorbed from 0, or unnoticeable quantities, to about a pint in a cubic foot. Its capacity for absorption varied according to quarry or the quality of the stone. Gneiss absorbed about as much as the poorer qualities of granite.

76. Limestone absorbed from two to

six pints in each cubic foot in most instances, although the best marbles absorbed less than a pint. Sandstones were found to absorb from about three to eight and a-half pints. Brown sandstones from New York and Connecticut absorbed about four pints, while the drab stones from Ohio take five or six pints.

77. The porosity of bricks from the same kiln, or stones from the same quarry, will vary greatly from each other. Where capillary attraction takes place in a comparatively dense material, it rises to a much greater height than in one which is very porous, as evaporation takes place more quickly in the latter instance. An open sandstone will be dry three feet above the level of complete saturation, while according to Eassie, dampness has been traced in a brick wall to the height of thirty-two feet. Materials which are permeable to water also allow air to pass through them in a greater or less degree. In this way, while protecting the building from water, ground-air is also being

excluded. Moisture which condenses on a cold wall, must not be mistaken for that which soaks through the wall. Thick and solid walls take so long to become warm that the moisture in the warm air of the house condenses on the surface. This usually evaporates quickly, but may be prevented by means of studding (Fig. 45), furring (Fig. 38 *c*) or hollow walls (Fig. 37 *b*).

78. *Animalculæ.* — The recent discovery of "rod-like animalculæ of the *genus bacilli*," in brickwork, by Mr. W. W. Goodrich,* and the discovery of similar germs by M. Parize,* in a different locality, make it the more important that impervious and solid material should be used in the construction of foundation-walls.

79. *Roman Walls.* — The Romans, whom no nations have surpassed in great constructions for convenience and comfort, if not for health, gave attention to proper methods of keeping their foundation-walls dry.

* *The American Architect and Building News*, August 4 and October 27, 1883.

Vitruvius,† writing about 25 B. C., says, in relation to the treatment of walls in a damp position: "First in apartments on a ground floor, a height of three feet above the pavement is to have its first coat of potsherd instead of sand, so that this part of the plastering may not be injured by the damp. But if a wall is liable to continual moisture, another thin wall should be carried up inside of it, as far within as the case will admit; and between the two walls a cavity is to be left lower than the level of the floor of the apartment, with openings for the air at the upper part; also openings must be left at the bottom, for if the damp does not evaporate through these holes above and below, it will extend to the new work. The work is then to be plastered with the potsherd mortar, made smooth and then polished with the last coat. If, however, there be not space enough for another wall, channels should nevertheless be made and

† Architecture of Vitruvius. English translation by Gwilt, 1860.

holes therefrom to the open air. Then tiles of the size of two feet are placed on one side of the channel, and on the other side piers are built of eight-inch bricks, on which the angles of two tiles may lie that they may not be distant more than one palm from each other. Over them other tiles with returning edges are fixed upright from the bottom to the top of the wall, the inner surface being carefully pitched over that they may resist the moisture; they are to have air-holes at the bottom and top above the vault. They are then to be whited over with lime and water, that the first coat may adhere to them, for, from the dryness they acquire in burning, they would neither take nor sustain this coat, but for the lime thus interposed, which joins and unites them."

Fig. 29 explains the manner of placing tile against the wall. This illustration, which is taken from Perrault's* French

* Les dix Architecture de Vitruve corrigez et traduits nouvellements en Francois avec des Notes et des Figures—Par. M. Perrault, a Paris, 1684.

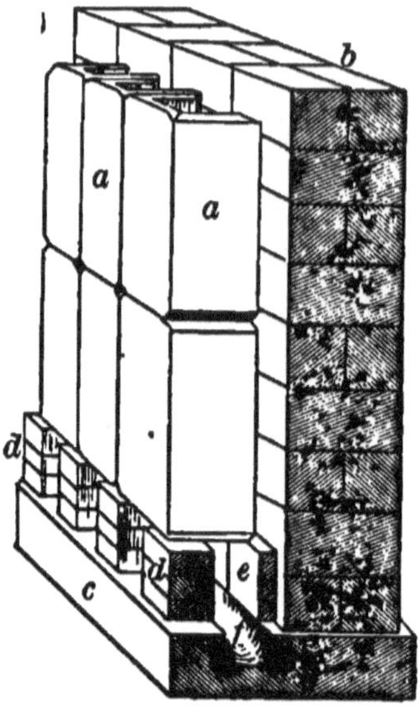


Fig. 29.—Roman Wall with Damp-Proof Tile.

*a*, Tile with returns; *b*, wall; *c*, drain; *d*, brick piers, with openings for ventilation between; *e*, plain tile; *f*, gutter in drain.

translation of Vitruvius, published in 1684, is republished, I think, for the first time. As the original illustrations of Vitruvius were lost, this shows Peraults interpretation of the text.

80. *Foundation in Roman Camp.*—According to Viollet Le Duc, the Romans observed great care in constructing their camps on hygienic principles, and where

the wall came against the side of a hill, as at the camp of the villa of Adrian at Tivoli, the walls were made double to avoid the dampness. Fig. 30 (after

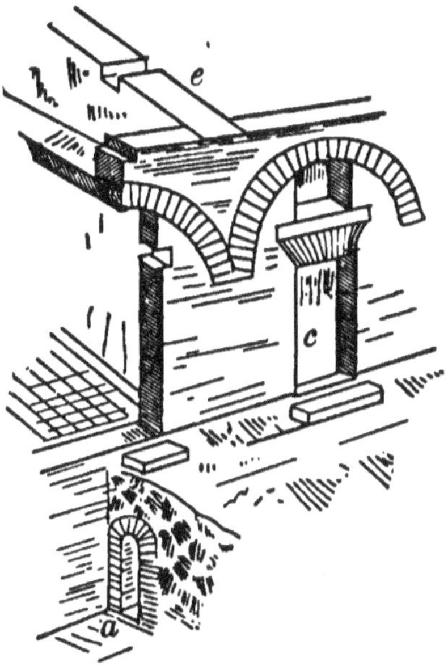


FIG. 30.—WALL AND FOUNDATION, ROMAN CAMP.

*a*, Concealed area or drain; *c*, entrance; *e*, roof-support.

Viollet Le Duc) is a view of a wall belonging to a Roman camp, in which a concealed area or intercepting drain is shown on the outside of the foundation-wall. Vitruvius tells us that following the Greeks the Romans built walls in three sections, the outside and inside walls being laid in large pieces and regular courses, and the space between was filled with either small stones thrown in loosely (when they would act as an intercepting drain), or the small stones were grouted with mortar, of which the

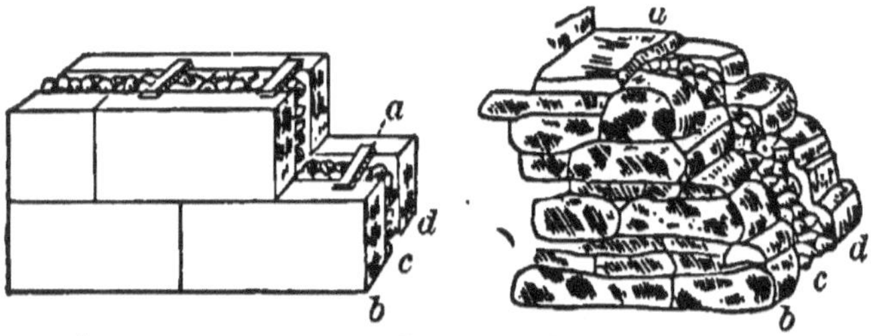


FIG. 31.—ROMAN WALLS.

*a*, Stone or iron-bound tie; *b*, outside; *c*, Loose stones or concrete; *d*, inside.

precise nature is not given. The outer walls were tied together by means of charred pieces of olive wood, or by bond stones in single pieces (διατονοι), running from one side to the other through the thickness of the wall (*a* Fig. 31). The latter method was borrowed from the Greeks. Sometimes iron clamps, "run with lead and duly preserving the middle space or cavity," were used to tie the interior and exterior walls together.

81. *Modern Walls.*—Whether it is advisable to build the wall of brick or stone, or whether the stone to be used is pervious or impervious, depends generally on the locality, as the material that is the most convenient and the cheapest is generally selected.

82. *Stone Formation.*—Hard, impervious stone makes the best foundation from a sanitary point of view. The formation of building stones is an interesting topic, to which but little space can be allowed in the present work. Geikie, in his recent work on mineralogy, gives an extended account of stone formations.

The earth's crust is composed of a few dominant substances or compounds. Oxygen forms about the half of all rocks. The most important compound is silicon dioxide or silica (Si O2), which composes more than half the known surface or crust of the earth, forming, as it does, most crystals and fragmental rocks. Silicic acid combines with earthy alkaline and metallic bases, and in this manner forms the different silicate salts. Calcium, principally in the form of carbonates of lime, alumina, magnesia, and iron (forming the principal natural pigment), with other substances, enter into the composition of building stones. Silicic, carbonic and sulphuric acids are the acids generally found in nature combined with the different bases, forming the silicates, carbonates and sulphates.

83. *Limestone*, usually calcium carbonate, ranges from a dirty grayish white stone to the finest marble. It is composed of granules more or less compact, and will absorb water freely. Marble and sandstones stand from 1000° to 1200° of heat before they yield, while granite only stands from 700° to 1000°.

84. *Clay slates*, being formed in layers, have cleavage, through the seams of which water would pass, although the openings might not be visible to the

naked eye; otherwise the stones are practicably impervious.

85. *Gneiss*, of which there are large deposits in this country, differs from granite principally in the foliation of the minerals—weak spots through which moisture may pass. Gneiss is composed of orthoclase, quartz and mica. The variation in this combination is great, and it is very difficult at times to distinguish it from granite.

86. *Quartz Rocks* differ from sandstone, in having the interstices between the grains of sand filled with a siliceous cementing material instead of having them open or filled with air. It makes a very compact foundation-stone.

87. *Granite* is a thorough admixture of felspar, mica and quartz, in granular and uniform sizes. According to Geikie, the quartz crystals in granite are filled with some liquid. This liquid, when subjected to a great heat, expands and bursts off fragments of the stone. In this way I have seen large granite piers completely destroyed.

88. *Sandstone* is generally composed of quartz grains, held together by some cementing material. These rocks are full of pores and interstices, from which the decomposable material has been taken away in solution, probably in the form of mud. Iron, in its different degrees of oxidation or hydration, colors sandstone red, brown, green or yellow. Sandstone is quarried in Connecticut, New York, Pennsylvania, Ohio, Virginia, and other parts of this country, and is used to a greater extent in this country, for a building-stone than any other kind. It is absorbent, and needs protection where it is used as a foundation in a damp location.

89. *Shale* is a hardened argillaceous rock, and varies on the one hand from clays that merge into slates, on the other from flagstones it merges into sandstones and limestones. In its best forms it is water-proof.

90. *Tests.*—Stones may be tested for cracks or fissures by striking them with a hammer, when the clear ring of the

perfect stone can be easily distinguished from the dull thud of the cracked stones. All cracks should be avoided, as they form passage-ways through which the moisture can enter.

91. *Stone Walls.*—The weak points in a stone wall are the joints and imperfect stones. If laid dry, or with the joints pointed only on the inside face, where they would be seen, the water would eventually wash the joints out, and the spring rains would fill the cellar of the building. Every stone should be well and solidly bedded in the mortar.

A common but objectionable method is that of hollowing out the side of the excavation so as to receive the large stones, when it would take the mason a little longer, or give him a little additional trouble to break the stone so as to make it of the proper size.

The projecting stones (Fig. 32 *b*) would catch all the water running down the side of the wall, and thence it would find its way through the joint to the inside of the house. The proper manner of

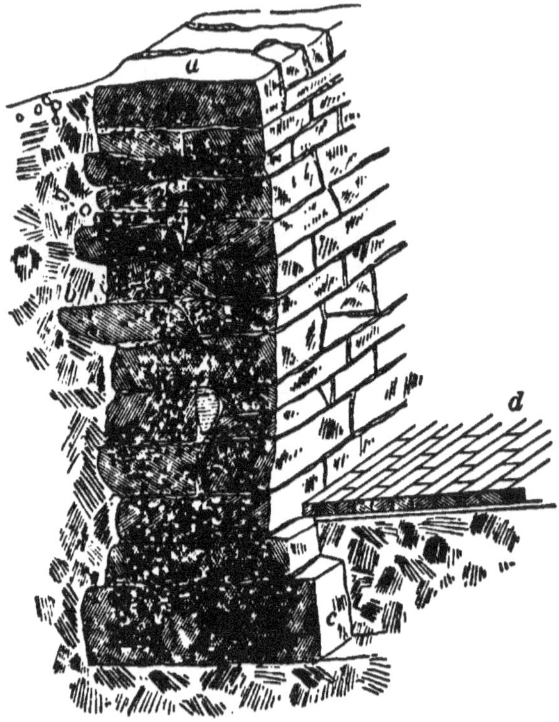


FIG. 32.—IMPERFECT STONE FOUNDATION-WALL.

*a*, Wall; *b*, projecting stone; *c*, stone footing; *d*, brick pavement.

building this kind of wall is to make the excavation eight or ten inches larger than is required for the foundation walls. After the wall is built the space should be filled to within a short distance of the surface with small broken stone or bricks.

FIG. 33.—PROPERLY-CONSTRUCTED STONE WALL,

*a*, Plaster; *b*, wall; *c*, concrete; *d*, asphalt-mastic; *e*, broken-stone footing; *f*, intercepting space filled with small or broken stone.

When treated in this way the space acts as an intercepting drain and catches the water from the subsoil and the surface, conveying it before it comes in contact with the wall into the drain beneath or on the outside of the footing. (Fig. 33.)

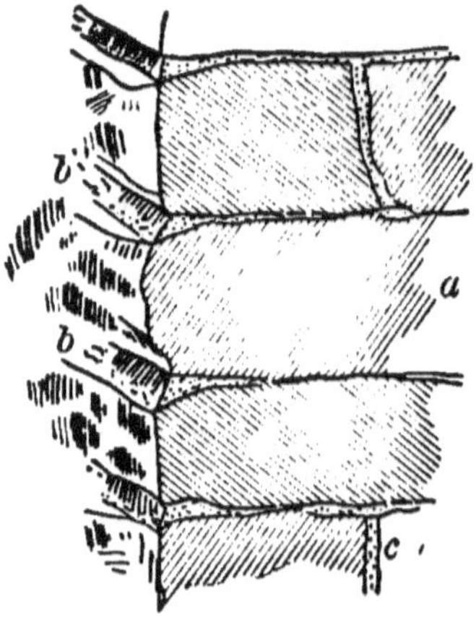


FIG. 34.—WEATHERED JOINT.

*a*, Wall; *b*, joint with downward inclination.

92. *Weather Joints.*—The outside face of the joint should always be beveled or weathered, *i. e.*, the point of the trowel must be run along the joint so as to give it a downward inclination of not less than 45° (Fig. 34). In this way all water that runs down on the face of the wall, instead of being impeded and soaking into the joint, will continue its course down to the drain and do no damage.

Where the stones composing the wall are impervious, a foundation of this kind will be found all that is needed, except in rare cases. Where the stones are absorbent, the walls must be protected in the manner brick walls are treated.

93. *Mortars.*—The combination of materials in which either stones or bricks are laid, or by which they are cemented together, forms an important item in a foundation wall. Mortar is a silicate of lime caused by the action of lime and sand under the influence of moisture. This acts slowly and gets harder year after year.

Lime mortar should never be used below ground, unless it is made from lime with hydraulic qualities.

Cement to an amount at least equal in bulk to the lime should be mixed with the mortar. A better or stronger mortar is made by having twice as much cement as there is lime. A pure cement mortar, no lime being used, is made by mixing one-third of cement and two thirds of sand.

94. *Lime.*—The quality of the lime is an important factor in making good mortar. Lime in powder, or in lumps that will not slake, is damaged. Lime that slakes quickly and leaves a hard core is underburnt, and when allowed to remain will swell in the wall and throw off the pointing or open the joint. Overburnt lime remains unslaked for a long time, and even when ground up slakes slowly. Lime is worthless in any one of the conditions named. Good lime slakes energetically, but never hardens under water. Plaster of Paris, which is of such service to the French builder, is sulphate of lime.

95. *Hydraulic Limes*—limes that set under water—are burned in the form of a dingy-colored lime, that slakes slowly after being ground up, and is called "ground lime." There is a black hydraulic lime, which slakes quickly, and is made from a pink marble. Mortar made from hydraulic limes may be used in foundation-walls.

96. *Sand* should be clean and sharp—

*i. e.*, prickly to the touch, and should leave no stain when compressed tightly in the hand.

97. *Manner of Making Mortar.*—A pen, with sides and bottom of boards, must be made to keep the mortar from the dirt. The lime is slaked in the pen by pouring in a barrel and a half of water to a barrel of lump lime. Water must not or should not be thrown in in driblets, as the partially slaked parts are thus chilled and form into lumps. The best plan is to measure it in, and pour it from a large cask; when done in any other way it is easy to get the lime paste either too thick or too thin. The pen may then be covered and left, as the vapor will cause the lime to slake more thoroughly than if it were left open, and it is advisable to let the paste stand as long as possible before it is mixed with the sand. Add to this paste, if the lime is good, two parts of sand to one part of lime paste. Lime mortar will stand for weeks (the majority of engineers think it is better for so standing)

FIG. 35.—FOUNDATION IN MARSHY GROUND.

*a*, Wall; *b*, cement-plaster; *c*, Terra-cotta guards; *d*, shed; *e*, water-inlet; *f*, broken-stone filling; *g*, asphalt-coating and damp-proof course; *h*, concrete; *i*, footing-stone; *k*, ties; *l*, piles; *m*, drain-tile; *n*, planking; *o*, marshy ground.

without setting, but it will get hard in a few hours after the addition of cement. A limited number of tests for cement will be given in connection with the description of concrete for footings.

98. *A Foundation in Marshy Ground.*—Where a marsh is of any depth, it is necessary, in the first place, to drive piles down until they rest upon or are driven into a solid or compact stratum, the heads of the piles being held in position by cross planks spiked or bolted to them. The whole should then be floored over with plank three or four inches thick (Fig. 31). Across this flooring, and over the line of piles, long footing stones may be placed. The wall, with its usual footing, is built above this course of stone. On the top of the first course of stone, and at least a foot above the flooring or woodwork, a line of two or three-inch drain tile is placed against the footings both inside and outside of the wall. Against the face of the wall on the outside ⊔ shaped terra-cotta pieces are so arranged as to make, practically, a hollow wall. It

is necessary to have openings in these terra-cotta tiles or guards (Fig. 35 *e*). Any water that passes through the open stone or intercepting drain will pass through the openings left for the purpose, down the side of the guards, into and off through the drain. These drains will keep the water below the cellar bottom, at the same time maintaining the water-level at least a foot above the wooden portion of the foundation. Wood submerged in water, the air never being allowed to come in contact with it, is practically imperishable.

A vitrified terra-cotta or stoneware shed is built into the wall above the guards to prevent water running down the side of the wall. The cellar bottom should then be covered with boards, if the marshy character of the ground makes it necessary, filled eight or ten inches with broken stone, and concreted above. The concrete should be coated with asphalt, and the asphalt should be carried up the inside face of the wall, over a coating of cement plastered over the stone to make an even surface.

99. *Footings.*—After the trenches have been excavated six or eight inches wider, inside and outside the building, than the wall above, and a solid bottom has been reached a foot or eighteen inches below the cellar floor, then the footing or first course or courses of the foundation wall are laid in these trenches.

The footings may be made from large stone selected for the purpose, long and wide, (Fig. 31*c*) or courses of brick are laid wider than the wall above, the number of courses and the width varying according to the superimposed structure (Fig 38). The course of brick in the bottom of the trenches being the widest, each course above is narrower for from five to eight courses, until the width of the wall is reached. The footing may be made by filling the trench with broken stone, rammed dry into place (Fig. 33*e*). The broken stone, in this case, forms an excellent drain encircling the building beneath the foundation walls. When there is a possibility of water collecting to a noticeable extent, these footings should have an outlet through either a stone or a tile drain to some low point, or should be connected with the sewer, as described in paragraphs 67–71.

From fear of settlements, some architects prefer to put their stone or concrete footings directly upon the solid bottom, and have the drain formed of broken stone or tile on the side instead of beneath the footing (Fig. 36*e*). Footing-stones, unless they are intended to act as drains, should be impervious, of good material, and have even beds. The trench should be hollowed out to receive any inequalites. If the stone does not set firmly it should be rammed down to a solid bearing, or, as Prof. T. M. Clark advises, sand may be packed around it, and when buckets of water are thrown on the sand it will be carried under the stone and fill all the crevices, and thus give the stone a solid bearing. Footings may be bedded solid in a thick layer of soft mortar with excellent effect.

On made or other ground, where the

natural foundation is poor, concrete footings may be put in extending over a large enough surface to carry the weight which is to be imposed.* Very little dampness will pass through good concrete foundations, Portland cement being practically impervious. And it is a good practice to make the footings and cellar bottom one continuous piece of concrete. (Fig. 43—*d.*)

100. *Cement.*—The quality of cement is the most important item in making concrete for footings or cellar floors, or in making mortar for jointing or plastering the face of the wall.

The architect has generally to rely on well known and reputable brands. Care must be taken to see that they have not been spoiled by water or dampness.

| * Loads that may be safely imposed on different natural foundations per sq. ft. | lbs. |
| --- | --- |
| Solid rock, from | 10,000 to 40,000 |
| Coarse sand, gravel, and clay, which must not be subjected to currents of running water | 4,000 " 6,000 |
| Piles, in made ground supported by friction | 3,500 " 4,000 |
| Piles, passing through made ground and solid ground | 8,000 " 10,000 |
| Piles, in solid ground | 30,000 "140,000 |

Cements and hydraulic limes are compounds (natural or artificial) of lime and other substances, chiefly alumina, which enables it to set harder and more rapidly under water than in its unmixed state. Natural cements are found in different parts of the country, and only need to be burned and pulverized, when they are ready for use. Artificial or Portland cement (so named because its color is similar to an English Portland stone), the strongest and best damp-proof cement, is prepared from a patented carbonate combined with chalk and limes.

To test cement in a rough way, as Professor Clark suggests, mix it with water and make small cakes, which are put away to dry until they are so dry that a small square stick, weighted by a brick, will make only a slight impression upon it. These cakes are immersed in water, and another set is made and left in the air.

101. After a lapse of twenty-four hours the piece left in the air should be hard, and should break with a clean frac-

ture instead of crumbling, while the cake left in the water should retain its shape and have increased in firmness. Machines are sometimes used for testing cement, the Government and private civil engineers making use of them in large works. Such machines are found necessary in order to prevent fraud and accidents in important works. For a de-cription of these machines see Abbott's "Testing Machines" in the Science Series.

102. *Concrete.*—For making concrete, the cement is mixed with two parts of clean, sharp sand, until neither a lump of cement nor patches of sand can be seen in the mass.

The mortar is called the matrix, while the broken stone or brick, gravel, bits of burnt clay, iron slag, or breeze, is called the aggregate. Enough water is mixed with the sand and cement to give it a pudding-like consistency. The aggregate must be wetted to wash off the light dust that might settle on it and prevent the matrix from adhering. In some instances the materials are mixed together separately and in a dry state, and then the whole mass is sprinkled with water.

When mixed in the proportion of one of cement, two of sand, five of broken stone or ballast, the concrete is wheeled to the desired point, and dumped from the height of two or three feet and quickly rammed.

103. *Laying Concrete.*—It is a mistaken idea that concrete should be thrown from an elevation of ten or twelve feet, as the large and heavy stones reach the bed first and the small ones remain at the top.

Concrete should never be laid more than twelve inches thick in one layer. When the first layer has hardened, it must be brushed off clean, wetted, and roughed up with a pick, when the second layer may be added. When laid in water, concrete can be laid in coarse cotton bags; the bags remain. Or, it may be deposited through shoots.

104. *Porous Stone Walls.*—Figs. 36 and 37 illustrate methods of treating

porous stone walls. Fig. 36 shows the outside faced with brick laid in asphalt; the asphalt continues through the wall, and forms a damp-proof course, while the cellar floor is paved with brick laid in asphalt.

FIG. 36.—POROUS-STONE WALLS.

*a* Brick laid in asphalt; *c*, wall; *b*, plaster; *d*, filling; *e*, drain; *f*, footing; *g*, asphalt.

Fig. 37 shows a double wall, in which the interior wall is built of brick. An air-space between the two walls prevents dampness from passing into the rooms.

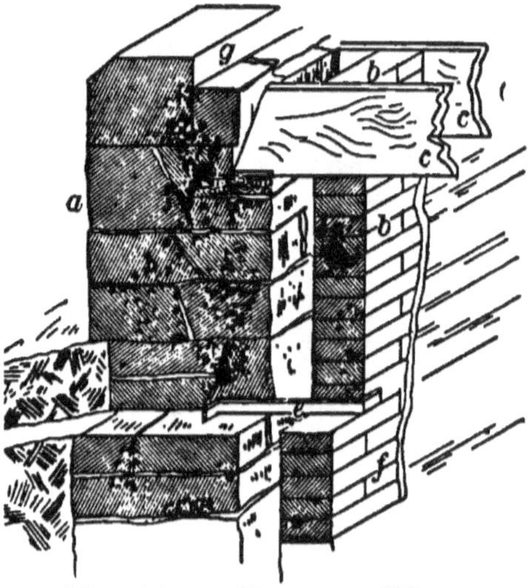

FIG. 37.—DOUBLE WALL.

*a*, Stone wall; *b*, brick wall; *c*, joist; *d*, projecting under-pinning; *e*, iron tie.

The underpinning, or part of the foundation-wall which is above ground, should overhang the wall below, so that water which trickles down the face of the wall, instead of continuing its course along the wall will run off into the earth or loose filling next to the foundation.

The manner of treating brick walls will answer equally well for stone walls that would absorb water.

### BRICK FOUNDATION-WALLS.

105. *Brick.*—Brick is used, I think, more extensively than stone for the purpose of building foundation-walls, and being porous, it readily absorbs moisture through its footings, or from the outside where it comes in contact with the earth. Bricks are usually found in three conditions as to hardness. Arch brick, as they form the arch in the kiln, come in direct contact with fire. These brick are hard, generally slightly vitrified or glazed, but they are usually distorted. When well shaped these arch bricks make the best foundations. The bricks farthest from the fire (Salmon) are imperfectly burned, and are worthless for foundations, as they absorb water and disintegrate easily. The medium bricks may be used when they are well burned.

106. *Brick Clays.*—There are three classes, although they verge from one

into the other, of brick earths or clays. (1) Earths, consisting of silica and alumina chiefly, with a small percentage of iron (hydrated oxide), lime, magnesia, manganese, etc. (2) Loams or sandy clays. (3) Marls or earths in which there is a considerable portion of lime. Where there is too much alumina, sand must be added, while loamy earth may need the addition of lime to act as a flux. The color of brick varies from the amount, degree of oxidation, or combination of the iron with other substances formed in the brick, and is therefore no criterion by which we can judge the quality of all brick. It is possible, when the character and color of brick (made from clay in a known locality) is familiar to the architect, for him to judge of their quality. Usually in this country light red or "salmon bricks" are inferior and unfit for foundations; dark red bricks are good and may be used in foundations, while bluish or greenish bricks are vitrified, and, if properly shaped, make the best foundations. Two bricks, when struck together, should have a clear ring, as a dull thud indicates either cracks or a want of proper adherence in the particles composing the bricks.

107. *Tempering.*—All clays should be "tempered"—that is, thoroughly mixed—and should go through a process of kneading similar to what they undergo in the pug-mill before being molded.

Where bricks are compressed by machinery, by great power, from clay in a dry or nearly dry state, they seem to disintegrate when exposed to the action of frost, and are not suitable for foundations.

108. *The Common Brick Foundation.*—The general manner of building brick foundations is very imperfect, as there must positively be no water or dampness in the ground beneath the surface, or it will show itself in the house. The excavation is made large enough, and only large enough, to receive the building, the soil in some cases being hollowed out to receive the footing, while the wall is built directly against the side

of the excavation. Just above the surface of the ground a double course of slate is introduced, the slate lapping over the joints in the course beneath it. This damp-proof course only protects the wall above the ground from dampness that would rise into it by capillary attraction. The plaster is prevented from showing dampness by furring the wall. Strips one by two inches or two by three inches are nailed to the wall at intervals of twelve or sixteen inches. On these strips the lathing and plastering are done, and an air-space left between the plaster and the wall (Fig. 38). These furring strips furnish flues and fuel for the flames and are the causes of many fires.

Wood in contact with a damp wall will soon decay, and, at best, a thin coat of plaster is but a poor protection against damp air carrying small particles of decayed vegetable matter suspended in it.

The cellar floor is covered with a coat of concrete one inch thick; on this is placed strips of cypress or white pine

FIG. 38.—COMMON OR IMPERFECT BRICK FOUNDATION-WALL.

*a*, Slate damp-proof course; *b*, brick wall; *c*, furring strip; *e*, flooring; *f*, concrete; *g*, floor joists.

two inches thick and three inches wide. The space between them is filled with a mastic composed of sand and pitch, the pitch being generally used instead of asphalt. The whole is rolled to an even surface, and the tongued and grooved flooring boards are nailed hard down against the mastic, or concrete, as it is called.

A covering of this kind is too thin to be of much value in protecting the building from moisture or ground-air.

109. *A Good Brick Foundation.*—To prevent the dampness and ground-air from entering our buildings, the footings may be made of broken stone, as described for stone-work, and the space next the wall, or between the wall and the embankment, should be filled, as mentioned before, with broken stone or small stones (Fig. 39). Where the flow of water is great, the stone-drain may be aided by running a line of small drain-tile in the stone-work, about an inch inside diameter. Authorities are divided in opinion in reference to the stability of broken stone beneath the foundation-wall. Where the natural foundation is good, I think there can be no doubt of its safety and utility. Where the natural

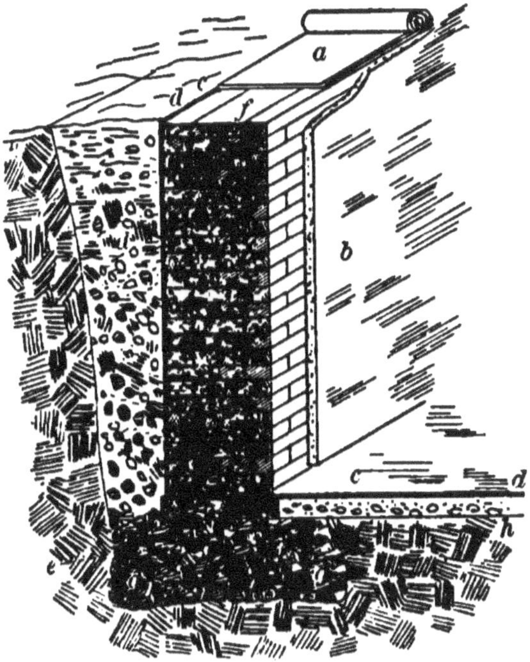


FIG. 39.—GOOD BRICK FOUNDATION.

*a*, Sheet-lead damp-proof course; *b*, plaster; *c*, asphalt coat; *d*, broken or small-stone filling; *e*, broken-stone footings; *g*, tile-drainage; *h*, concrete.

foundation is bad or the weight of the superstructure great, then large stones or concrete must be used, as described in other parts of this work.

The wall above the footing-course and the surface of the cellar bottom is protected by a coating of asphalt, forming, as it does, a continuous coating from the surface of the ground along the face of and through the wall, and over the cellar floor.

110. *Hollow Walls.*—One of the simplest, and, at the same time, one of the most effective methods of protecting the interior of the building from dampness, is to build the foundation-walls hollow or with an air space between them.

A hollow wall is, in fact, two walls built up separately, with an air-space not less than three inches wide between them.

When less than three inches, this space is liable to be clogged up by droppings of mortar, from the wall above. Mr. E. S. Philbrick finds a 4-inch space necessary, and openings left at the bottom to clean out the droppings.

The two walls must be tied or bound together, to prevent a tendency they would have to spread out or double up from the weight of the wall above; for this purpose different forms of iron and brick ties are used.

111. *Different Iron Ties.*—The simplest iron tie is made by bending a wrought-iron bar a quarter of an inch thick two inches from each end (Fig. 40). When placed in the wall flat any water from the outside is liable to run along the top of the iron and into the inside wall. When the tie is placed on edge the possibility of water passing from one wall to the other is diminished by the difference between the breadth and thickness of the tie.

Iron ties are sometimes twisted or bent in the form of a U or V where they span the open space between the walls, so that any water from the outside would drop off in the hollow space between the walls.

Instead of simply bending the ends of the ties, plates may be bolted or riveted to them. In this way a larger surface of brick work is brought within the influence of the tie. The plates may be made ornamental where they show, either

FIG. 40.—DIFFERENT IRON TIES.

*a*, Inside wall; *d*, outside wall; *b*, broken-stone footing; *c*, tile-drain; *e*, plain tie (flat); *f*, plain tie, on edge; *g*, twisted tie; *h*, V-shaped tie; *i*, plate ends; *k*, tie running through the wall; *l*, inclined tie; *m*, stone damp-proof course; *n*, asphalt coating and damp-proof course; *o*, concrete.

from necessity or design, on the outside of the wall (Fig. 40).

112. *Preservation of Ties.*—To prevent the iron from being destroyed by oxidation, it must be coated with paint, red-lead being the best, pitch, asphalt, or zinc (galvanized). Pitched or galvanized ties are the most commonly used.

Ties treated by the Bower-Barff process would be well protected, but I am not familiar with the cost of this process.

In ordinary cases ties should be put in every sixth course, and between two and three feet apart. Instead of placing the ties above each other in the different rows, they should be so placed as to bring the tie in one row midway between the ties in the rows above and below it. Where walls are to be subjected to the jarring effects of railroads or machinery, the number of ties should be increased, being put in every three or four courses, and every eighteen inches to two feet apart.

113. *Brick Ties.*—Common brick, and

FIG. 41.—BRICK TIES.

*a*, Wall; *b* and *c*, common bricks; *d*, curved brick; *e*, wedge-shaped brick with holes in center; *f*, same, running only partly through the wall; *g*, footing; *h*, concrete; *i*, damp-proof course of terra-cotta.

brick molded for the purpose, are used for tying hollow walls together. Even when common bricks are coated with an impervious material, such as gas-tar, pitch or asphalt, and it is prevented from conveying water by capillary attraction, nevertheless water will pass along the surface from one wall to the other.

114. A brick adapted to the purpose was designed by Jennings, of England. These bricks are designed so as to project from the outside of one wall to the inside of the other, or so as to reach only within four inches of the face of each wall, and thus not to interfere with the bond on the outside.

The ends are wedge-shaped, which gives them a better hold on the wall, and they have holes in the center through which the greater part of the water would trickle into the hollow space below (Fig. 41—*e*).

The best brick tie was designed by the same party; it is so shaped that the inside is one or two courses higher than the outside end. By this form water is prevented from passing along the top of the brick (Fig. 41—*d*). All such bricks should be made from vitrified or glazed terra-cotta or stoneware.

115. *Thickness of Hollow Walls.*—The two walls composing a hollow wall may be of equal thickness, when it will give the most substantial foundation.

The inside wall may be thicker than the outside when the smaller amount of brick-work will be exposed to dampness, or the thick wall may be built on the outside, with no advantage unless the whole lot is to be utilized in the building above the foundation (Fig. 37).

116. *Dwarf-Wall.*—A hollow wall is sometimes made by building a small dwarf-wall three or four inches from the main wall, with bricks projecting at regular intervals and resting against the main wall. The projecting bricks on one side, and the filling packed on the other side give the wall stability.

If the ends of the bricks which come in contact with the main wall are protected by an impervious coating in a

FIG. 42.—DWARF-WALL.

*a*, Terra-cotta damp-proof course; *b*, main-wall; *c*, dwarf-wall; *d*, concrete; *e*, tile-drain.

wall of this kind, it will form an effectual barrier against dampness on the face of the wall where the ground comes in contact with it (Fig. 42).

## IV.

### MISCELLANEOUS.

117. *Impervious Coatings.*—Brick and stone walls that would absorb water require an impervious coating on the outside, in addition to the broken-stone

filling, where there is a sign of moisture in the soil or sub-soil.

118. For this purpose numerous expedients have been tried; enameled or glazed brick, with joints laid in asphalt and fine sand, or with joints laid in Portland cement mixed with same quantity of clean fine sand; an excellent method, but too expensive for ordinary use. Bricks have been made for the purpose with glass in their center, and asphalt or other bituminous material mixed with sand has been molded in the form of damp-proof brick.

Glazed terra-cotta slabs, worked in so as to protect the wall, were introduced into England by Follet in 1869. Facing walls with non-absorbing materials is common in Germany, density and non-absorbency going hand in hand.

119. An effectual impervious facing is given to the wall by using Taylor's or Doulton's damp-proof courses, made of vitrified or glazed stoneware (Fig. 43). The tile must be laid with the vitrified surface on the outside, next to the embankment.

The tiles may be tied into the wall either by iron clamps (asphalted or galvanized), or the tiles may be built into the wall so that every other tile will have its narrow edge outside and run back its full depth into the wall. The holes which run through the tile must not open on the outside, as they would form conduits through which water could run into the wall. Joints should be made with asphalt or cement and fine sand.

120. *Cheap Facings.* — The cheap methods of protecting the face of a wall where it is below the surface are by facing it with slate laid in cement-mortar, plastering it with cement-mortar, or coating it with asphalt-mastic, gas-tar, or pitch. Asphalt requires a smooth surface, so the wall should be plastered first with cement-mortar. Gas-tar and pitch must be applied in a boiling state, and the bricks must be perfectly dry, and it is better to have them heated. Cement may crack and slate become disjointed, while bitumens, mineral or organic, have

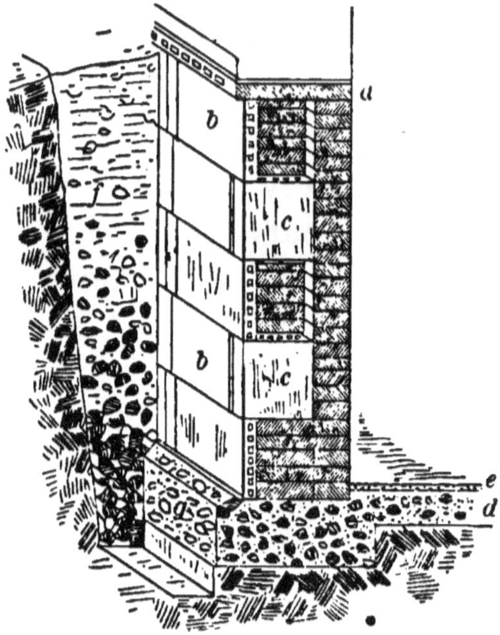


Fig. 43.—Tile Facing.

*a*, Tile damp-proof course; *b*, tile-facing; *c*, bond-tile; *d*, concrete footing and cellar-bottom; *e*, cement coat; *f*, broken-stone filling.

a certain elasticity to them which makes them valuable.

121. *Damp-Proof Courses.*—Impervious courses are introduced into brick or other absorbent walls to prevent moisture that would come through the footings or other unprotected parts of

the wall which are below the surface of the ground from rising into the wall above. Such courses are called damp-proof, and are inserted just above the ground line, and when properly laid they are an effective barrier to water rising by capillary attraction.

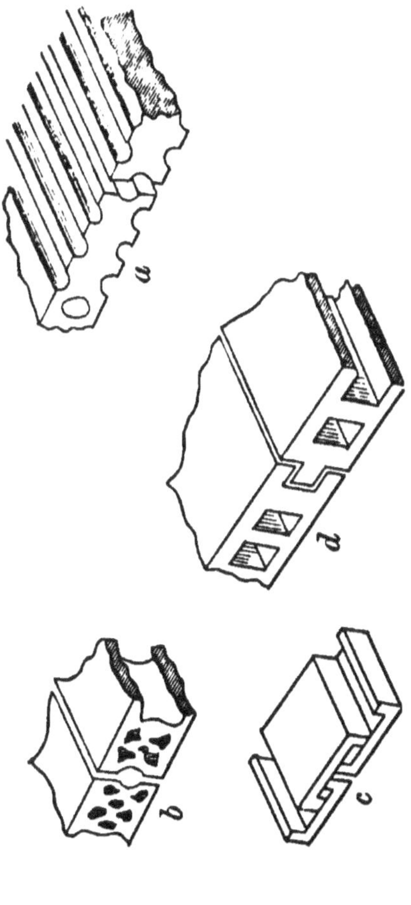

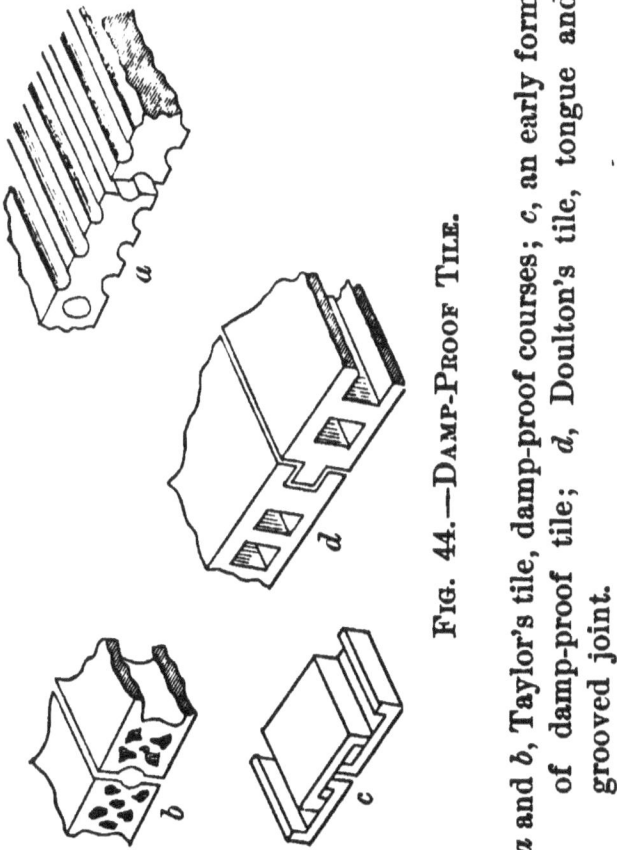

FIG. 44.—DAMP-PROOF TILE.

*a* and *b*, Taylor's tile, damp-proof courses; *c*, an early form of damp-proof tile; *d*, Doulton's tile, tongue and grooved joint.

122. *Tile Damp-Proof Courses.*—The Taylor and Doulton damp-proof courses (Fig. 44), made hollow and of highly-vitrified stoneware, are used in England.

The Doulton tile has a tongue which fits into a corresponding groove in the tile next to it in the course. These tiles are made so as to cover the entire width of the wall, and must be jointed with Portland cement or asphalt and sand.

The holes which run through all these tiles may be utilized for ventilation.

123. *Sheet-Lead*, cut the width or thickness of the wall, weight five or six pounds per superficial foot, makes an excellent but expensive damp-proof course (Fig. 39—*a*). At the joint the lead should be carried down in a vertical joint and lapped. A solder joint might break, either from expansion or contraction or by settlement in the masonry.

*Hot Asphalt* (Fig. 39—*d*) mixed with sand, one-half to three-quarters of an inch thick, makes a good damp-proof course. Bitumen damp-proof courses are now manufactured of any width or length.

Portland cement, mixed with sand, about one-half of an inch thick, is sometimes used as a damp-proof, but it is liable to crack, and thus prove ineffectual.

124. *Slate* (Fig. 38—*a*) laid in cement, two courses deep, in which the top course covers the joint of the bottom course, is the damp-proof course used almost universally in this country and in England. Slate, when carefully bedded in cement, will not break nnless there is an unusual settlement in the foundation. When the slate laps the joints properly, it makes a damp-proof course, which serves its purpose well, and will not be superseded except in expensive buildings.

125. *Interior Coating.*—Treatment of the interior face of the wall, except in connection with the treatment of the exterior, or to protect plastering or frescoing from stain, amounts to nothing more than covering the evil up to keep it out of sight. The damp wall, with its evil effects are unremoved.

Many quack remedies have been devised, each of which, according to the *inventor*, is perfect. The interior surface (as a sure cure) has been covered with tin foil or oil cloths. Glass veneer has been tried in France and Germany, the glass being held in position by litharge plaster or cement. The walls have been coated with paints of many compounds, alums, soaps, silicates, and cements or asphalt, all have been tried. The water is merely dammed up, the wall is wet, and the water will eventually find its way through.

When it is desirable to treat the wall, so as to insure the protection of frescoing, the wall may be studded (Fig. 45–*g*).

126. *Studding Walls.*—The studs being independent of the wall (two by four inches in section) are stiff enough to plaster on. The space between studs should be filled in solid at the top and bottom with two or three courses of brick to prevent the spread of fire.

In this way an air-space is formed that will add materially to the comfort of the house. This plan may be adopted in connection with some of the treatments

FIG. 45.

*a*, Wall; *e*, damp-proof course; *f*, water-shed; *g*, studding; *h*, lath and plaster.

mentioned for the outside of the wall with advantage.

127. *Water Sheds.*—Sometimes water sheds are placed beneath the surface to carry off the water that runs down the side of the building, or falls directly over the shed (Fig. 45). The shed is composed of a course of bricks laid against the building near the surface, and running off into the ground at an angle of forty-five degrees, and extending to a distance of two or three feet from the building line. The shed must be covered with some impervious material, such as

slate laid in cement or a coating of cement or asphalt and sand, the slate cover being probably the best. The first course must be laid in a groove, cut or left in the wall for that purpose. The shed should be built on concrete or boards (in some cases), where the foundation is bad and liable to settle.

128. Water sheds are sometimes advantageously placed above ground; in that case they take the form of either an asphalt or concrete pavement. Where the ground is sandy or gravelly, a shed of this kind is useful, as it carries the surface water well away from the building before it soaks into the ground.

129. *Areas.*—One of the most common and effectual methods of protecting an outside wall from dampness or moisture is by means of an area. Areas may be concealed, being covered with earth and connected with the main wall at the top, or they may be open—that is, entirely independent of the building.

130. *French Treatment.* —Viollet le Duc describes a method for checking a

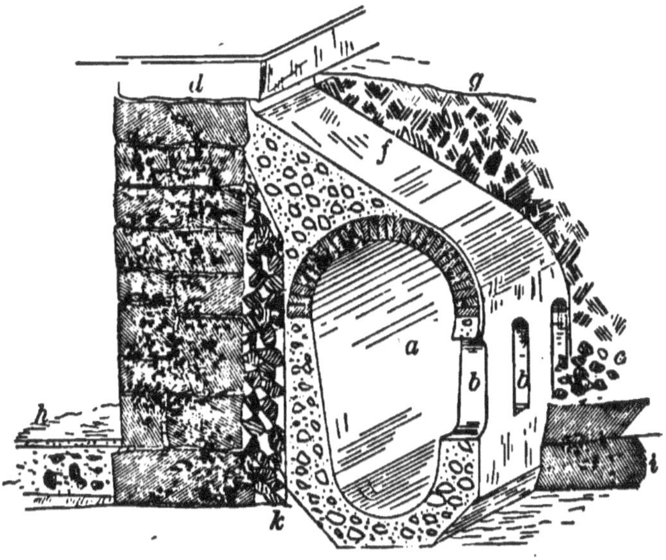


FIG. 46—FRENCH AREA OR INTERCEPTING-DRAIN.

*a*, Area; *b*, openings; *c*, stone filling; *d*, impervious stone damp-proof course; *f*, concrete water-shed; *g*, earth filling; *h*, concrete; *i*, impervious stratum; *k*, loose stone between area and building.

flow of water that would otherwise run against and into the foundation, by means of a concealed passageway or area (Fig. 46). This area is intended to intercept any water that might flow along a rocky or other impervious stratum toward the building.

The passageway is made of concrete, in the form of an egg-shaped sewer, with an arch of stone or brick for its crown.

131. A slanting surface of concrete extends from the building line to the outside edge of the sewer. In this way a water shed is formed that would throw the surface water away from the building. On the side of this conduit, away from the building, there is a row of long, narrow openings or slits, against which stones are piled. The water that runs along the impervious stratum is intercepted, and enters the passage by the open slits, and may be conveyed away as described on a previous page.

132. These passageways may be entirely independent of the wall, the space between being filled with broken stone, or the foundation-wall may be used as the interior wall of the conduit (Fig. 47). The first method would protect the building better than the latter, while the latter would be the cheaper.

133. *Concealed Area.*—A convenient concealed area may be built after the

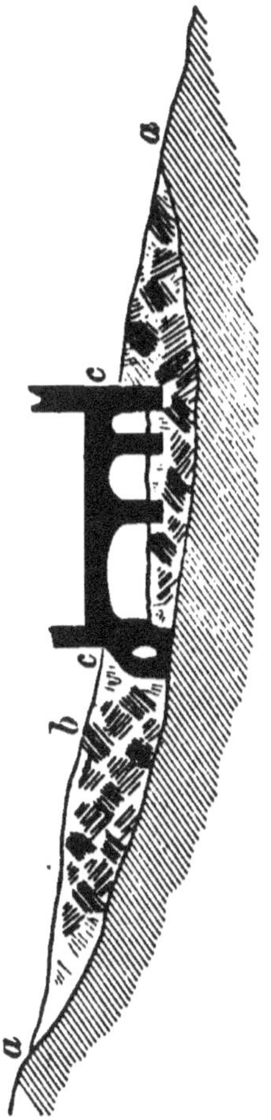

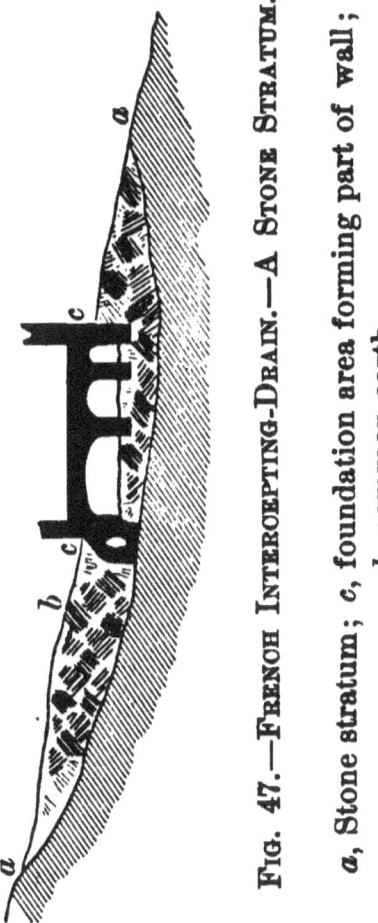

FIG. 47.—FRENCH INTERCEPTING-DRAIN.—A STONE STRATUM.

*a*, Stone stratum; *c*, foundation area forming part of wall; *b*, common earth.

main wall is carried up. The concrete foundation should be extended about three feet beyond the foundation wall, and on this a nine-inch, or one-brick thick wall may be carried up to within a

FIG. 48.—CONCEALED AREA.

*a*, Water-shed, cemented; *b*, main wall; *c*, damp-proof stone; *d*, area-wall; *i*, tile-drain; *f*, perforated tile; *g*, area bottom, iron grating; *h*, area; *l*, concrete floor and footings; *k*, joists; *m*, double floors, tar-paper between them; *n*, surface of the ground.

short distance of the surface, when an arch should be sprung from it to the foundation wall. The top of the brick forming this arch may be cut or molded so as to form a water shed, and throw the surface water away from the building (Fig. 48).

The top of this brick-work must be covered with asphalt or cement and sand.

134. The open space between the two walls should be at least large enough to allow a man to enter, as the additional expense would be small, and the convenience at some time might be great.

The concrete must be graded toward the center, and coated with cement or asphalt. Beneath the center or lowest point in the drain a tile drain must be laid to carry off all the water that might come through the area wall.

135. It is always advisable to provide for the passage of water through an area wall, instead of damming it up by the wall.

The perforations in the damp-proof

course answer both to let water and air through the wall. Although there is very little danger of dirt getting into a concealed area, still it is well to cover the gutter with a cast-iron perforated grating, as it will prevent rats and mice from entering the drain. Concealed areas must have two or more openings into the outside air, so that a circulation of pure air may take place through the area, and through which any ground air can pass into the pure air on the outside of the building.

136. *Manhole.*—Concealed areas should have a manhole through which they can be inspected from time to time.

Areas are sometimes built in the form of a number of small arches. A very thin area wall may be built in this way, as the arch abuts against the wall of the building. Although this form of area is used in England, it is objectionable on account of the corner, which would be hard to clean. (Fig. 49.)

137. *Open Areas.*—An open area is nothing more than a low retaining wall

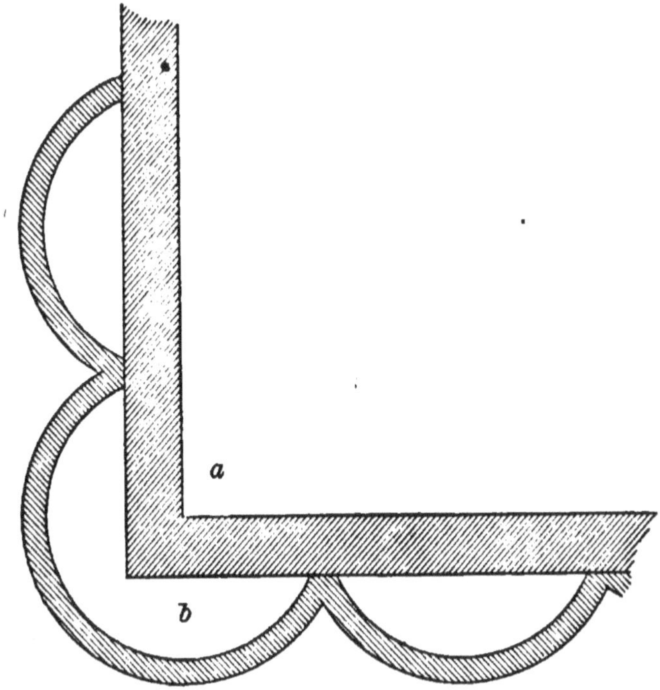


FIG. 49.—PLAN OF AREA FORMED BY ARCHES.

*a*, Wall; *b*, area.

built against the side of the cut, two or more feet from the foundation-wall. An area of this kind forms a complete protection for the wall against moisture or dampness that might come from the bank of earth piled against the wall. An open grating may be placed over an open area of this kind by letting it fit in rebates made in the coping of the dwarf wall,

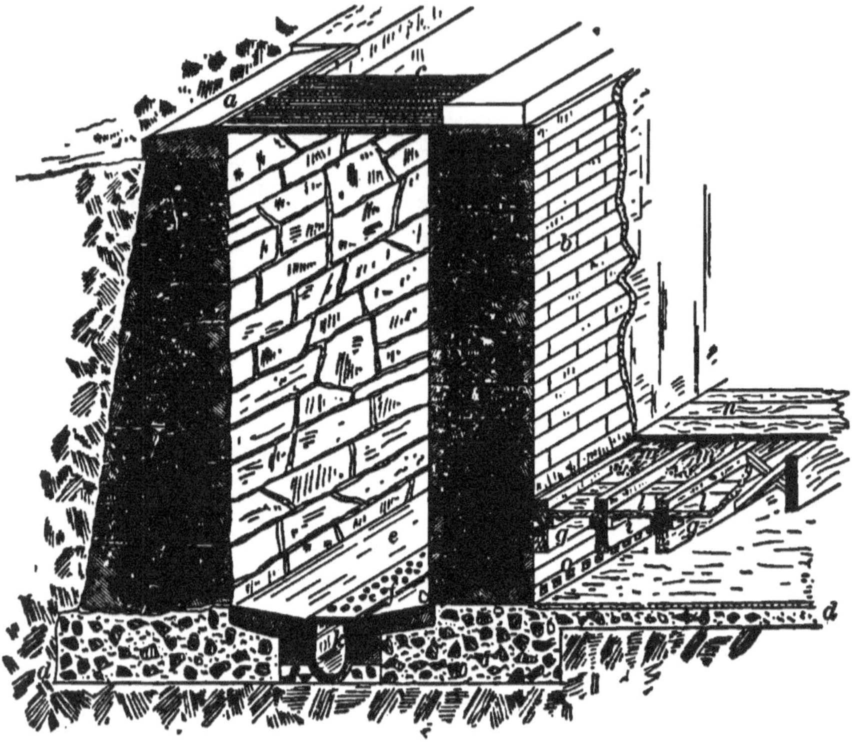


FIG. 50.—OPEN AREA.

*a*, Coping; *b*, main wall; *c*, grating; *d*, concrete; *e*, flagstones; *f*, perforated iron cover; *g*, joists, ventilator beneath; *i*, straps on joists; *k*, horseshoe-tile; *l*, false floor; *m*, asbestos filling or packing; *n*, floor.

and in the stone belt of the foundation wall (Fig. 50).

138. Open areas are always to be preferred to concealed ones. They are open to the purifying action of light and air. Being constantly under the eye, they will be kept clean by a careful housewife. Concealed areas, being out of sight, are generally out of mind. Vermin delight to congregate in dark places when they dare not show themselves in places open to the light of day.

139. *Cellar Floors.* — The common method of laying a brick pavement in sand or lime mortar offers no impediment to either water or air. Bricks laid in cement mortar, and covered with a coating of the same, are better, while bricks laid on in asphalt mastic, and covered with it, form an excellent cellar bottom. The bricks protect the asphalt, while the asphalt prevents either moisture or impure air from entering the building. (See Fig. 36—*a*.)

140. All cellar floors should be concreted with a coating of concrete from

three to six inches thick, according to circumstances.

The concrete must then be finished with a coating of cement mortar or asphalt mastic. The concrete should be made in the same manner as that described for footing courses. (See Fig. 50-*d.*)

141. *Asphalt.*—Contractors sometimes attempt and sometimes commit an imposition by using vegetable pitch, gas tar combined with chalk, and other compositions, in the place of asphalt.

Asphalt is a mineral pitch, found practically, although not chemically, pure in nature. It has been used from the earliest times for the purpose of protecting substances or bodies from air and water. The Egyptians, probably more than three thousand years ago, used asphalt to protect their mummies.

142. A limestone impregnated with asphalt is used in France instead of the simple substance. In Washington City, where the number of square yards of asphalt pavement exceeds that of all the

other cities in the world combined, the following formula has been found to make the best mastic:

| | |
|---|---|
| *Asphaltic cement (refined Trinidad asphalt, 100 parts; petroleum oil, 20 parts) | 15 to 18 |
| Limestone powder | 15 to 17 |
| Sand | 70 to 65 |
| | 100 to 100 |

This coating neither cracks in cold weather (10°), nor becomes soft in hot weather (160° Fah.) in the sun.

Asphalt is found to be far superior in durability to any of its substitutes.

143. *Floors.*—Where it is necessary or desirable to have the cellar floor boarded over, as is the case when this part of the building is to be used as a basement, the flooring is sometimes laid directly on sleepers imbeded in the concrete. (See Fig. 38–*g*.) The only feature which a floor of this kind has to recommend it is, that rats and mice cannot find a playground beneath it. Where the joists are raised above the concrete covering,

* Engineer's Report of the District of Columbia.

and perforated damp-proof tiles (Fig. 50-*g*) are built into the wall, a circulation of air will take place beneath the floor.

All basement floors are best laid double, with two thicknesses of tarred paper between them.

When the joists are not bedded in the concrete, small strips may be nailed to the side of the joists one-third of the way from the top, and pieces of board sawed so as to rest on them. In this manner a false floor is formed, with a space two or three inches deep between it and the top floor.

144. *Mineral Wool.* — This space should be filled with mineral wool, a material prepared by passing superheated steam through ordinary iron slag. It has the appearance of white wool, and weighs very little in comparison with its bulk.

It is said, on good authority, to be death to insects that enter it. This material is fire-proof, and a non-conductor. Mineral wool is liable to decay where it becomes wet, and as it absorbs and retains moisture, it should not be used in damp places. These spaces may be filled with asbestos, as it forms a good non-conducting material.

145. *Roof Water.*—It seems scarcely necessary to say that water from the roof should not be allowed to fall from the eaves to the ground, as it would run into the ground and against the foundation wall. It should always be carried off in properly constructed gutters and down spouts.

146. The down-spouts should, when there is a drainage system, be carried into the disconnecting-trap between the drainage and sewerage systems. When the down-spout connects directly with the sewer it should have a deep trap, and be carried up, of cast iron, with lead-calked joints, or of wrought iron with screw joints, to a point above the roof of the building. With the ordinary tin or zinc-coated iron down-spouts there are always outlets for sewer air, and, as the trap may lose its seal by evaporation, the

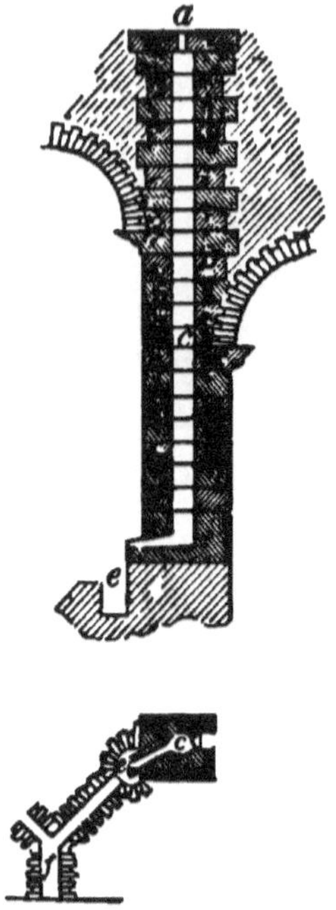


FIG. 51.—SECTION AND PLAN OF DOWN-SPOUTS IN COLISEUM.

*a*, Capstone; *c*, down-spout or drain; *e*, catch-basin; *f*, sewer.

sewer air would be drawn into the house through the windows.

147. *The Coliseum* affords an interesting example of the methods employed by the ancients to drain off rain water.

According to Edward Cresy,* fifty-six of

* "Encyclopedia of Engineering," Edward Cresy.

these drain pipes, or, more properly, down-spouts, remain in the thickness of the Coliseum wall. These conduits, about twelve inches in diameter, were hollowed from the middle of large masses of freestone twenty inches high, and alternately jutting into the wall, so as to bond with the other parts of the masonry (Fig. 51).

149. "The upper end of each stone is cut in the form of an inverted cone, while the bottom is level." In this way water was prevented from entering the masonry through the joint. The top of these conduits was capped by a large stone; decidedly dish-shaped on top, with a small hole two inches in diameter passing through it into the conduit.

150. In this way all the seats and galleries were drained. The water passed off into gutters, and thence through small brick or stone drains into the large sewers for which Rome is famous.

# VALUABLE

# SCIENTIFIC BOOKS

PUBLISHED BY

## D. VAN NOSTRAND,

23 MURRAY STREET AND 27 WARREN STREET, N. Y.

ADAMS (J. W.) Sewers and Drains for Populous Districts. Embracing Rules and Formulas for the dimensions and construction of works of Sanitary Engineers. Second edition. 8vo, cloth .......... $2 50

ALEXANDER (J. H.) Universal Dictionary of Weights and Measures, Ancient and Modern, reduced to the standards of the United States of America. New edition, enlarged. 8vo, cloth .......... 3 50

ATWOOD (GEO.) Practical Blow-Pipe Assaying. 12mo, cloth, illustrated .......... 2 00

AUCHINCLOSS (W. S.) Link and Valve Motions Simplified. Illustrated with 37 wood-cuts and 21 lithographic plates, together with a Travel Scale and numerous useful tables. 8vo, cloth .......... 3 00

AXON (W. E. A.) The Mechanic's Friend: a Collection of Receipts and Practical Suggestions Relating to Aquaria—Bronzing—Cements—Drawing—Dyes—Electricity—Gilding—Glass-working—Glues—Horology—Lacquers—Locomotives—Magnetism—Metal-working—Modelling—Photography—Pyrotechny—Railways—Solders—Steam-Engine—Telegraphy—Taxidermy—Varnishes—Waterproofing, and Miscellaneous Tools, Instruments, Machines, and Processes connected with the Chemical and Mechanic Arts. With numerous diagrams and wood-cuts. Fancy cloth .......... 1 50

BACON (F. W.) A Treatise on the Richards Steam-Engine Indicator, with directions for its use. By Charles T. Porter. Revised, with notes and large additions as developed by American practice; with an appendix containing useful formulæ and rules for engineers. Illustrated. Third edition. 12mo, cloth .......... 1 00

BARBA (J.) The Use of Steel for Constructive Purposes; Method of Working, Applying, and Testing Plates and Brass. With a Preface by A. L. Holley, C.E. 12mo, cloth. $1 50

BARNES (Lt. Com. J. S., U. S. N.) Submarine Warfare, offensive and defensive, including a discussion of the offensive Torpedo System, its effects upon Iron-Clad Ship Systems and influence upon future naval wars. With twenty lithographic plates and many wood-cuts. 8vo, cloth............ 5 00

BEILSTEIN (F.) An Introduction to Qualitative Chemical Analysis, translated by I. J. Osbun. 12mo, cloth.......... 75

BENET (Gen. S. V., U. S. A.) Electro-Ballistic Machines, and the Schultz Chronoscope. Illustrated. Second edition, 4to, cloth.................................................... 3 00

BLAKE (W. P.) Report upon the Precious Metals: Being Statistical Notices of the principal Gold and Silver producing regions of the World, represented at the Paris Universal Exposition. 8vo, cloth....................................... 2 00

—— Ceramic Art. A Report on Pottery, Porcelain, Tiles, Terra Cotta, and Brick. 8vo, cloth......................... 2 00

BOW (R. H.) A Treatise on Bracing, with its application to Bridges and other Structures of Wood or Iron. 156 illustrations. 8vo, cloth.............................. ......... 1 50

BOWSER (Prof. E. A.) An Elementary Treatise on Analytic Geometry, embracing Plane Geometry, and an Introduction to Geometry of three Dimensions. 12mo, cloth....... 1 75

—— An Elementary Treatise on the Differential and Integral Calculus. With numerous examples. 12mo, cloth......... 2 25

BURGH (N. P.) Modern Marine Engineering, applied to Paddle and Screw Propulsion. Consisting of 36 colored plates, 259 practical wood-cut illustrations, and 403 pages of descriptive matter, the whole being an exposition of the present practice of James Watt & Co., J. & G. Rennie, R. Napier & Sons, and other celebrated firms. Thick 4to vol., cloth ........... .................. ........................10 00
Half morocco......... ......................................15 00

BURT (W. A.) Key to the Solar Compass, and Surveyor's Companion; comprising all the rules necessary for use in the field; also description of the Linear Surveys and Public Land System of the United States, Notes on the Barometer, suggestions for an outfit for a survey of four months, etc. Second edition. Pocket-book form, tuck............ 2 50

BUTLER (Capt. J. S., U. S. A.) Projectiles and Rifled Cannon. A Critical Discussion of the Principal Systems of Rifling and Projectiles, with Practical Suggestions for their Improvement, as embraced in a Report to the Chief of Ordnance, U. S. A. 36 plates. 4to, cloth...................... 6 00

CAIN (Prof. WM.) A Practical Treatise on Voussoir and Solid and Braced Arches. 16mo, cloth extra .................... $1 75

CALDWELL (Prof. GEO. C.) and BRENEMAN (Prof. A. A.) Manual of Introductory Chemical Practice, for the use of Students in Colleges and Normal and High Schools. Third edition, revised and corrected. 8vo, cloth, illustrated. New and enlarged edition .................... 1 50

CAMPIN (FRANCIS). On the Construction of Iron Roofs. 8vo, with plates, cloth .................... 2 00

CHAUVENET (Prof. W.) New method of correcting Lunar Distances, and improved method of finding the error and rate of a chronometer, by equal altitudes. 8vo, cloth ...... 2 00

CHURCH (JOHN A.) Notes of a Metallurgical Journey in Europe. 8vo, cloth .................... 2 00

CLARK (D. KINNEAR, C.E.) Fuel: Its Combustion and Economy, consisting of Abridgments of Treatise on the Combustion of Coal and the Prevention of Smoke, by C. W. Williams; and the Economy of Fuel, by T. S. Prideaux. With extensive additions on recent practice in the Combustion and Economy of Fuel: Coal, Coke, Wood, Peat, Petroleum, etc. 12mo, cloth .................... 1 50

—— A Manual of Rules, Tables, and Data for Mechanical Engineers. Based on the most recent investigations. Illustrated with numerous diagrams. 1,012 pages. 8vo, cloth ... 7 50
Half morocco .................... 10 00

CLARK (Lt. LEWIS, U. S. N.) Theoretical Navigation and Nautical Astronomy. Illustrated with 41 wood-cuts. 8vo, cloth .................... 1 50

CLARKE (T. C.) Description of the Iron Railway Bridge over the Mississippi River at Quincy, Illinois. Illustrated with 21 lithographed plans. 4to, cloth .................... 7 50

CLEVENGER (S. R.) A Treatise on the Method of Government Surveying, as prescribed by the U. S. Congress and Commissioner of the General Land Office, with complete Mathematical, Astronomical, and Practical Instructions for the use of the United States Surveyors in the field. 16mo, morocco .................... 2 50

COFFIN (Prof J. H. C.) Navigation and Nautical Astronomy. Prepared for the use of the U. S. Naval Academy. Sixth edition. 52 wood-cut illustrations. 12mo, cloth ...... 3 50

COLBURN (ZERAH). The Gas-Works of London. 12mo, boards .................... 60

COLLINS (JAS. E.) The Private Book of Useful Alloys and Memoranda for Goldsmiths, Jewellers, etc. 18mo, cloth ... 50

CORNWALL (Prof. H. B.) Manual of Blow-Pipe Analysis, Qualitative and Quantitative, with a Complete System of Descriptive Mineralogy. 8vo, cloth, with many illustrations. N. Y., 1882 .......... $2 50

CRAIG (B. F.) Weights and Measures. An account of the Decimal System, with Tables of Conversion for Commercial and Scientific Uses. Square 32mo, limp cloth .......... 50

CRAIG (Prof. THOS.) Elements of the Mathematical Theory of Fluid Motion. 16mo, cloth .......... 1 25

DAVIS (C. B.) and RAE (F. B.) Hand-Book of Electrical Diagrams and Connections. Illustrated with 32 full-page illustrations. Second edition. Oblong 8vo, cloth extra .......... 2 00

DIEDRICH (JOHN). The Theory of Strains: a Compendium for the Calculation and Construction of Bridges, Roofs, and Cranes. Illustrated by numerous plates and diagrams. 8vo, cloth .......... 5 00

DIXON (D. B.) The Machinist's and Steam-Engineer's Practical Calculator. A Compilation of useful Rules, and Problems Arithmetically Solved, together with General Information applicable to Shop-Tools, Mill-Gearing, Pulleys and Shafts, Steam-Boilers and Engines. Embracing Valuable Tables, and Instruction in Screw-cutting, Valve and Link Motion, etc. 16mo, full morocco, pocket form ...(In press)

DODD (GEO.) Dictionary of Manufactures, Mining, Machinery, and the Industrial Arts. 12mo, cloth .......... 1 50

DOUGLASS (Prof. S. H.) and PRESCOTT (Prof. A. B.) Qualitative Chemical Analysis. A Guide in the Practical Study of Chemistry, and in the Work of Analysis. Third edition. 8vo, cloth .......... 3 50

DUBOIS (A. J.) The New Method of Graphical Statics. With 60 illustrations. 8vo, cloth .......... 1 50

EASSIE (P. B.) Wood and its Uses. A Hand-Book for the use of Contractors, Builders, Architects, Engineers, and Timber Merchants. Upwards of 250 illustrations. 8vo, cloth. 1 50

EDDY (Prof. H. T.) Researches in Graphical Statics, embracing New Constructions in Graphical Statics, a New General Method in Graphical Statics, and the Theory of Internal Stress in Graphical Statics. 8vo, cloth .......... 1 50

ELIOT (Prof. C. W.) and STORER (Prof. F. H.) A Compendious Manual of Qualitative Chemical Analysis. Revised with the co-operation of the authors. By Prof. William R. Nichols. Illustrated. 12mo, cloth .......... 1 50

ELLIOT (Maj. GEO. H., U. S. E.) European Light-House Systems. Being a Report of a Tour of Inspection made in 1873. 51 engravings and 21 wood-cuts. 8vo, cloth .......... 5 00

ENGINEERING FACTS AND FIGURES. An Annual Register of Progress in Mechanical Engineering and Construction for the years 1863-64-65-66-67-68. Fully illustrated. 6 vols. 18mo, cloth (each volume sold separately), per vol. .... $2 50

FANNING (J. T.) A Practical Treatise of Water-Supply Engineering. Relating to the Hydrology, Hydrodynamics, and Practical Construction of Water-Works in North America. Third edition. With numerous tables and 180 illustrations. 650 pages. 8vo, cloth .... 5 00

FISKE (BRADLEY A., U. S. N.) Electricity in Theory and Practice. 8vo, cloth .... 2 50

FOSTER (Gen. J. G., U. S. A.) Submarine Blasting in Boston Harbor, Massachusetts. Removal of Tower and Corwin Rocks. Illustrated with seven plates. 4to, cloth .... 3 50

FOYE (Prof. J. C.) Chemical Problems. With brief Statements of the Principles involved. Second edition, revised and enlarged. 16mo, boards .... 50

FRANCIS (JAS. B., C. E.) Lowell Hydraulic Experiments: Being a selection from Experiments on Hydraulic Motors, on the Flow of Water over Weirs, in Open Canals of Uniform Rectangular Section, and through submerged Orifices and diverging Tubes. Made at Lowell, Massachusetts. Fourth edition, revised and enlarged, with many new experiments, and illustrated with twenty-three copperplate engravings. 4to, cloth .... 15 00

FREE-HAND DRAWING. A Guide to Ornamental Figure and Landscape Drawing. By an Art Student. 18mo, boards .... 50

GILLMORE (Gen. Q. A.) Treatise on Limes, Hydraulic Cements, and Mortars. Papers on Practical Engineering, U. S. Engineer Department, No. 9, containing Reports of numerous Experiments conducted in New York City during the years 1858 to 1861, inclusive. With numerous illustrations. 8vo, cloth .... 4 00

—— Practical Treatise on the Construction of Roads, Streets, and Pavements. With 70 illustrations. 12mo, cloth .... 2 00

—— Report on Strength of the Building Stones in the United States, etc. 8vo, illustrated, cloth .... 1 50

—— Coignet Beton and other Artificial Stone. 9 plates, views, etc. 8vo, cloth .... 2 50

GOODEVE (T. M.) A Text-Book on the Steam-Engine. 143 illustrations. 12mo, cloth .... 2 00

GORDON (J. E. H.) Four Lectures on Static Induction. 12mo, cloth .... 80

GRUNER (M. L.) The Manufacture of Steel. Translated from the French, by Lenox Smith, with an appendix on the Bessemer process in the United States, by the translator. Illustrated. 8vo, cloth.......................................$3 50

HALF-HOURS WITH MODERN SCIENTISTS. Lectures and Essays. By Professors Huxley, Barker, Stirling, Cope, Tyndall, Wallace, Roscoe, Huggins, Lockyer, Young, Mayer, and Reed. Being the University Series bound up. With a general introduction by Noah Porter, President of Yale College. 2 vols. 12mo, cloth, illustrated .............. 2 50

HAMILTON (W. G.) Useful Information for Railway Men. Sixth edition, revised and enlarged. 562 pages, pocket form. Morocco, gilt.............................................. 2 00

HARRISON (W. B.) The Mechanic's Tool Book, with Practical Rules and Suggestions for Use of Machinists, Iron-Workers, and others. Illustrated with 44 engravings. 12mo, cloth.............................................. 1 50

HASKINS (C. H.) The Galvanometer and its Uses. A Manual for Electricians and Students. Second edition. 12mo, morocco.............................................. 1 50

HENRICI (OLAUS). Skeleton Structures, especially in their application to the Building of Steel and Iron Bridges. With folding plates and diagrams. 8vo, cloth.................... 1 50

HEWSON (WM.) Principles and Practice of Embanking Lands from River Floods, as applied to the Levees of the Mississippi. 8vo, cloth.................................... 2 00

HOLLEY (ALEX. L.) A Treatise on Ordnance and Armor, embracing descriptions, discussions, and professional opinions concerning the materials, fabrication, requirements, capabilities, and endurance of European and American Guns, for Naval, Sea-Coast, and Iron-Clad Warfare, and their Rifling, Projectiles, and Breech-Loading; also, results of experiments against armor, from official records, with an appendix referring to Gun-Cotton, Hooped Guns, etc., etc. 948 pages, 493 engravings, and 147 Tables of Results, etc. 8vo, half roan ..............................................10 00

—— Railway Practice American and European Railway Practice in the economical Generation of Steam, including the Materials and Construction of Coal-burning Boilers, Combustion, the Variable Blast, Vaporization, Circulation, Superheating, Supplying and Heating Feed-water, etc., and the Adaptation of Wood and Coke-burning Engines to Coal-burning; and in Permanent Way, including Road-bed, Sleepers, Rails, Joint-fastenings, Street Railways, etc., etc. With 77 lithographed plates. Folio, cloth..................12 00

HOWARD (C. R.) Earthwork Mensuration on the Basis of the Prismoidal Formulæ. Containing simple and labor-saving method of obtaining Prismoidal Contents directly from End Areas. Illustrated by Examples, and accompanied by Plain Rules for Practical Uses. Illustrated. 8vo, cloth ........ $1 50

INDUCTION-COILS. How Made and How Used. 63 illustrations. 16mo, boards ........ 50

ISHERWOOD (B. F.) Engineering Precedents for Steam Machinery. Arranged in the most practical and useful manner for Engineers. With illustrations. Two volumes in one. 8vo, cloth ........ 2 50

JANNETTAZ (EDWARD). A Guide to the Determination of Rocks: being an Introduction to Lithology. Translated from the French by G. W. Plympton, Professor of Physical Science at Brooklyn Polytechnic Institute. 12mo, cloth.... 1 50

JEFFERS (Capt. W. N., U. S. N.) Nautical Surveying. Illustrated with 9 copperplates and 31 wood-cut illustrations. 8vo, cloth ........ 5 00

JONES (H. CHAPMAN). Text-Book of Experimental Organic Chemistry for Students. 18mo, cloth ........ 1 00

JOYNSON (F. H.) The Metals used in Construction: Iron, Steel, Bessemer Metal, etc., etc. Illustrated. 12mo, cloth. 75

—— Designing and Construction of Machine Gearing. Illustrated. 8vo, cloth ........ 2 00

KANSAS CITY BRIDGE (THE). With an account of the Regimen of the Missouri River, and a description of the methods used for Founding in that River. By O. Chanute, Chief-Engineer, and George Morrison, Assistant-Engineer. Illustrated with five lithographic views and twelve plates of plans. 4to, cloth ........ 6 00

KING (W. H.) Lessons and Practical Notes on Steam, the Steam-Engine, Propellers, etc., etc., for young Marine Engineers, Students, and others. Revised by Chief-Engineer J. W. King, U. S. Navy. Nineteenth edition, enlarged. 8vo, cloth ........ 2 00

KIRKWOOD (JAS. P.) Report on the Filtration of River Waters for the supply of Cities, as practised in Europe, made to the Board of Water Commissioners of the City of St. Louis. Illustrated by 30 double-plate engravings. 4to, cloth ........ 15 00

LARRABEE (C. S.) Cipher and Secret Letter and Telegraphic Code, with Hogg's Improvements. The most perfect secret code ever invented or discovered. Impossible to read without the key. 18mo, cloth ........ 1 00

LOCK (C. G.), WIGNER (G. W.), and HARLAND (R. H.) Sugar Growing and Refining. Treatise on the Culture of Sugar-Yielding Plants, and the Manufacture and Refining of Cane, Beet, and other sugars. 8vo, cloth, illustrated ......12 00

LOCKWOOD (THOS. D.) Electricity, Magnetism, and Electro-Telegraphy. A Practical Guide for Students, Operators, and Inspectors. 8vo, cloth.............................$2 50

LORING (A. E.) A Hand-Book on the Electro-Magnetic Telegraph. Paper boards.............................50
Cloth.............................75
Morocco.............................1 00

MacCORD (Prof. C. W.) A Practical Treatise on the Slide-Valve by Eccentrics, examining by methods the action of the Eccentric upon the Slide-Valve, and explaining the practical processes of laying out the movements, adapting the valve for its various duties in the steam-engine. Second edition Illustrated. 4to, cloth.............................2 50

McCULLOCH (Prof. R. S.) Elementary Treatise on the Mechanical Theory of Heat, and its application to Air and Steam Engines. 8vo, cloth.............................3 50

MERRILL (Col. WM. E., U. S. A.) Iron Truss Bridges for Railroads. The method of calculating strains in Trusses, with a careful comparison of the most prominent Trusses, in reference to economy in combination, etc., etc. Illustrated. 4to, cloth.............................5 00

MICHAELIS (Capt. O. E., U. S. A.) The Le Boulenge Chronograph, with three lithograph folding plates of illustrations. 4to, cloth.............................3 00

MICHIE (Prof. P. S.) Elements of Wave Motion relating to Sound and Light. Text-Book for the U.S. Military Academy. 8vo, cloth, illustrated.............................5 00

MINIFIE (WM.) Mechanical Drawing. A Text-Book of Geometrical Drawing for the use of Mechanics and Schools, in which the Definitions and Rules of Geometry are familiarly explained; the Practical Problems are arranged, from the most simple to the more complex, and in their description technicalities are avoided as much as possible. With illustrations for Drawing Plans, Sections, and Elevations of Railways and Machinery; an Introduction to Isometrical Drawing, and an Essay on Linear Perspective and Shadows. Illustrated with over 200 diagrams engraved on steel. Ninth edition. With an Appendix on the Theory and Application of Colors. 8vo, cloth.............................4 00

"It is the best work on Drawing that we have ever seen, and is especially a text-book of Geometrical Drawing for the use of Mechanics and Schools. No young Mechanic, such as a Machinist, Engineer, Cabinet-maker, Millwright, or Carpenter, should be without it."—*Scientific American.*

—— Geometrical Drawing. Abridged from the octavo edition, for the use of schools. Illustrated with forty-eight steel plates. Fifth edition. 12mo, cloth.............................2 00

MODERN METEOROLOGY. A Series of Six Lectures, delivered under the auspices of the Meteorological Society in 1878. Illustrated. 12mo, cloth.... ....... .. .... .....$1 50

MORRIS (E.) Easy Rules for the Measurement of Earthworks, by Means of the Prismoidal Formula. 78 illustrations. 8vo, cloth........................................ 1 50

MOTT (H. A., Jr.) A Practical Treatise on Chemistry (Qualitative and Quantitative Analysis), Stoichiometry, Blow-pipe Analysis, Mineralogy, Assaving, Pharmaceutical Preparations, Human Secretions, Specific Gravities, Weights and Measures, etc., etc., etc. New edition, 1883. 650 pages. 8vo, cloth...................... ......... ............. 4 00

NAQUET (A.) Legal Chemistry. A Guide to the Detection of Poisons, Falsification of Writings, Adulteration of Alimentary and Pharmaceutical Substances, Analysis of Ashes, and examination of Hair, Coins, Arms, and Stains, as applied to Chemical Jurisprudence, for the use of Chemists, Physicians, Lawyers, Pharmacists, and Experts. Translated, with additions, including a list of books and Memoirs on Toxicology, etc., from the French. By J. P. Battershall, Ph.D., with a preface by C. F. Chandler, Ph.D., M.D., LL.D. 12mo, cloth....... .... .. ...................... 2 00

NOBLE (W. H.) Useful Tables. Pocket form, cloth......... 50

NUGENT (E.) Treatise on Optics; or, Light and Sight, theoretically and practically treated, with the application to Fine Art and Industrial Pursuits. With 103 illustrations. 12mo, cloth..... ........................................ 1 50

PEIRCE (B.) System of Analytic Mechanics. 4to, cloth.....10 00

PLANE TABLE (THE). Its Uses in Topographical Surveying. From the Papers of the U. S. Coast Survey. Illustrated. 8vo, cloth........................................ 2 00

"This work gives a description of the Plane Table employed at the U. S. Coast Survey office, and the manner of using it."

PLATTNER. Manual of Qualitative and Quantitative Analysis with the Blow-Pipe. From the last German edition, revised and enlarged. By Prof. Th. Richter, of the Royal Saxon Mining Academy. Translated by Prof. H. B. Cornwall, assisted by John H. Caswell. Illustrated with 87 woodcuts and one lithographic plate. Fourth edition, revised, 560 pages. 8vo, cloth................................ 5 00

PLYMPTON (Prof. GEO. W.) The Blow-Pipe. A Guide to its use in the Determination of Salts and Minerals. Compiled from various sources. 12mo, cloth.......................... 1 50

—— The Aneroid Barometer: Its Construction and Use. Compiled from several sources. 16mo, boards, illustrated, 50
Morocco .................................................. 1 00

PLYMPTON (Prof. GEO. W.) The Star-Finder, or Planisphere, with Movable Horizon. Printed in colors on fine card-board, and in accordance with Proctor's Star Atlas... $1 00

POCKET LOGARITHMS, to Four Places of Decimals, including Logarithms of Numbers, and Logarithmic Sines and Tangents to Single Minutes. To which is added a Table of Natural Sines, Tangents, and Co-Tangents. 16mo, boards, 50
Morocco........................................................ 1 00

POOK (S. M.) Method of Comparing the Lines and Draughting Vessels propelled by Sail or Steam. Including a chapter on Laying-off on the Mould-Loft Floor. 1 vol. 8vo, with illustrations, cloth........................................ 5 00

POPE (F. L.) Modern Practice of the Electric Telegraph. A Hand-Book for Electricians and Operators. Eleventh edition, revised and enlarged, and fully illustrated. 8vo, cloth. 2 00

PRESCOTT (Prof. A. B.) Outlines of Proximate Organic Analysis, for the Identification, Separation, and Quantitative Determination of the more commonly occurring Organic Compounds. 12mo, cloth.................................... 1 75

—— Chemical Examination of Alcoholic Liquors. A Manual of the Constituents of the Distilled Spirits and Fermented Liquors of Commerce, and their Qualitative and Quantitative Determinations. 12mo, cloth........................... 1 50

—— First Book in Qualitative Chemistry. Second edition. 12mo, cloth................................................ 1 50

PYNCHON (Prof. T. R.) Introduction to Chemical Physics, designed for the use of Academies, Colleges, and High-Schools. Illustrated with numerous engravings, and containing copious experiments with directions for preparing them. New edition, revised and enlarged, and illustrated by 269 illustrations on wood. Crown 8vo, cloth............ 3 00

RAMMELSBERG (C. F.) Guide to a Course of Quantitative Chemical Analysis, especially of Minerals and Furnace Products. Illustrated by Examples. Translated by J. Towler, M.D. 8vo, cloth........................................ 2 25

RANDALL (P. M.) Quartz Operator's Hand-Book. New edition, revised and enlarged, fully illustrated. 12mo, cloth... 2 00

RANKINE (W. J. M.) Applied Mechanics, comprising Principles of Statics, Cinematics, and Dynamics, and Theory of Structures, Mechanism, and Machines. Crown 8vo, cloth. Tenth edition. London..... ..... .............. 5 00

—— A Manual of the Steam-Engine and other Prime Movers, with numerous tables and illustrations. Crown 8vo, cloth. Tenth edition. London, 1882.............................. 5 00

—— A Selection from the Miscellaneous Scientific Papers of, with Memoir by P. G. Tait, and edited by W. J. Millar, C.E. 8vo, cloth. London, 1880................................10 00

RANKINE (W. J. M.) A Manual of Machinery and Mill-work. Fourth edition. Crown 8vo. London, 1881 ........ .... $5 00

—— Civil Engineering, comprising Engineering Surveys, Earthwork, Foundations, Masonry, Carpentry, Metal-works, Roads, Railways, Canals, Rivers, Water-works, Harbors, etc., with numerous tables and illustrations. Fourteenth edition, revised by E. F. Bamber, C.E. 8vo. London, 1883.......................................... 6 50

—— Useful Rules and Tables for Architects, Builders, Carpenters, Coachbuilders, Engineers, Founders, Mechanics, Shipbuilders, Surveyors, Typefounders, Wheelwrights, etc. Sixth edition. Crown 8vo, cloth. London, 1883...... 4 00

—— and BAMBER (E. F.) A Mechanical Text-Book; or, Introduction to the Study of Mechanics and Engineering. 8vo, cloth. London, 1875 ................................ 3 50

RICE (Prof. J. M.) and JOHNSON (Prof. W. W.) On a New Method of Obtaining the Differentials of Functions, with especial reference to the Newtonian Conception of Rates or Velocities. 12mo, paper ....... .. ...................... 50

ROGERS (Prof. H. D.) The Geology of Pennsylvania. A Government Survey, with a General View of the Geology of the United States, Essays on the Coal Formation and its Fossils, and a description of the Coal Fields of North America and Great Britain. Illustrated with Plates and Engravings in the text. 3 vols. 4to, cloth, with Portfolio of Maps. ....30 00

ROEBLING (J. A.) Long and Short Span Railway Bridges. Illustrated with large copperplate engravings of plans and views. Imperial folio, cloth ............................ 25 00

ROSE (JOSHUA, M.E.) The Pattern-Maker's Assistant, embracing Lathe Work, Branch Work Core Work, Sweep Work, and Practical Gear Constructions, the Preparation and Use of Tools, together with a large collection of useful and valuable Tables. Third edition. Illustrated with 250 engravings. 8vo, cloth ............ ............... ....... 2 50

SABINE (ROBERT). History and Progress of the Electric Telegraph, with descriptions of some of the apparatus. Second edition, with additions, 12mo, cloth .... .................. 1 25

SAELTZER (ALEX.) Treatise on Acoustics in connection with Ventilation. 12mo, cloth .................................. 1 00

SCHUMANN (F.) A Manual of Heating and Ventilation in its Practical Application for the use of Engineers and Architects, embracing a series of Tables and Formulæ for dimensions of heating, flow and return pipes for steam and hot-water boilers, flues, etc., etc. 12mo. Illustrated. Full roan ... ... ....... .... ................................ 1 50

—— Formulas and Tables for Architects and Engineers in calculating the strains and capacity of structures in Iron and Wood. 12mo, morocco, tucks ...................... 2 50

SAWYER (W. E.) Electric-Lighting by Incandescence, and its Application to Interior Illumination. A Practical Treatise. With 96 illustrations. Third edition. 8vo, cloth. $2 50

SCRIBNER (J. M.) Engineers' and Mechanics' Companion, comprising United States Weights and Measures, Mensuration of Superfices and Solids, Tables of Squares and Cubes, Square and Cube Roots, Circumference and Areas of Circles, the Mechanical Powers, Centres of Gravity, Gravitation of Bodies, Pendulums, Specific Gravity of Bodies, Strength, Weight, and Crush of Materials, Water-Wheels, Hydrostatics, Hydraulics, Statics, Centres of Percussion and Gyration, Friction Heat, Tables of the Weight of Metals, Scantling, etc., Steam and the Steam-Engine. Nineteenth edition, revised, 16mo, full morocco............ 1 50

—— Engineers', Contractors', and Surveyors' Pocket Table-Book. Comprising Logarithms of Numbers, Logarithmic Sines and Tangents, Natural Sines and Natural Tangents, the Traverse Table, and a full and complete set of Excavation and Embankment Tables, together with numerous other valuable tables for Engineers, etc. Eleventh edition, revised, 16mo, full morocco ............................ 1 50

SHELLEN (Dr. H.) Dynamo-Electric Machines. Translated, with much new matter on American practice, and many illustrations which now appear for the first time in print. 8vo, cloth, New York............................(In press)

SHOCK (Chief-Eng. W. H.) Steam-Boilers: their Design, Construction, and Management. 450 pages text. Illustrated with 150 wood-cuts and 36 full-page plates (several double). Quarto. Illustrated. Half morocco......................15 00

SHUNK (W. F.) The Field Engineer. A handy book of practice in the Survey, Location, and Track-work of Railroads, containing a large collection of Rules and Tables, original and selected, applicable to both the Standard and Narrow Gauge, and prepared with special reference to the wants of the young Engineer. Third edition. 12mo, morocco, tucks.................................................. 2 50

SHIELDS (J. E.) Notes on Engineering Construction. Embracing Discussions of the Principles involved, and Descriptions of the Material employed in Tunnelling, Bridging, Canal and Road Building, etc., etc. 12mo, cloth........... 1 50

SHREVE (S. H.) A Treatise on the Strength of Bridges and Roofs. Comprising the determination of Algebraic formulas for strains in Horizontal, Inclined or Rafter, Triangular, Bowstring, Lenticular, and other Trusses, from fixed and moving loads, with practical applications and examples, for the use of Students and Engineers. 87 wood-cut illustrations. Third edition. 8vo, cloth......................... 3 50

SIMMS (F. W.) A Treatise on the Principles and Practice of Levelling, showing its application to purposes of Railway Engineering and the Construction of Roads, etc. Revised and corrected, with the addition of Mr. Laws's Practical Examples for setting out Railway Curves. Illustrated. 8vo, cloth .......................................... $2 50

STILLMAN (PAUL.) Steam-Engine Indicator, and the Improved Manometer Steam and Vacuum Gauges—their Utility and Application. New edition. 12mo, flexible cloth....... 1 00

STONEY (B. D.) The Theory of Strains in Girders and similar structures, with observations on the application of Theory to Practice, and Tables of Strength and other properties of Materials. New and revised edition, enlarged. Royal 8vo, 664 pages. Complete in one volume. 8vo, cloth........... 12 50

STUART (CHAS. B., U. S. N.) The Naval Dry Docks of the United States. Illustrated with 24 engravings on steel. Fourth edition, cloth .......................................... 6 00

—— The Civil and Military Engineers of America. With 9 finely executed portraits of eminent engineers, and illustrated by engravings of some of the most important works constructed in America. 8vo, cloth....................... 5 00

STUART (B.) How to Become a Successful Engineer. Being Hints to Youths intending to adopt the Profession. Sixth edition. 12mo, boards .................................. 50

SWEET (S. H.) Special Report on Coal, showing its Distribution, Classification, and Cost delivered over different routes to various points in the State of New York and the principal cities on the Atlantic Coast. With maps. 8vo, cloth .......................................... 3 00

TEXT-BOOK (A) ON SURVEYING, Projections, and Portable Instruments, for the Use of the Cadet Midshipmen at the U. S. Naval Academy. Nine lithographed plates and several wood-cuts. 8vo, cloth .............................. 2 00

TONER (J. M.) Dictionary of Elevations and Climatic Register of the United States. Containing, in addition to Elevations, the Latitude, Mean Annual Temperature, and the total Annual Rain-fall of many localities; with a brief introduction on the Orographic and Physical Peculiarities of North America. 8vo, cloth.............................. 3 75

TUCKER (Dr. J. H.) A Manual of Sugar Analysis, including the Applications in General of Analytical Methods to the Sugar Industry. With an Introduction on the Chemistry of Cane Sugar, Dextrose, Levulose, and Milk Sugar. 8vo, cloth, illustrated.............................. 3 50

TUNNER (P.) A Treatise on Roll-Turning for the Manufacture of Iron Translated and adapted by John B. Pearse, of the Pennsylvania Steel-Works, with numerous engravings, wood-cuts, and folio atlas of plates................. 10 00

VAN WAGENEN (T. F.)' Manual of Hydraulic Mining, for the use of the Practical Miner. 12mo, cloth.............. $1 00

WALKER (W. H.) Screw Propulsion. Notes on Screw Propulsion: Its Rise and History. 8vo, cloth.................. 75

WANKLYN (J. A.) A Practical Treatise on the Examination of Milk and its Derivatives, Cream, Butter, and Cheese. 12mo, cloth................ ...... ........................ 1 00

WATT (ALEX.) Electro-Metallurgy, Practically Treated. Sixth edition, with considerable additions. 12mo, cloth.... 1 00

WEISBACH (JULIUS). A Manual of Theoretical Mechanics. Translated from the fourth augmented and improved German edition, with an introduction to the Calculus, by Eckley B. Coxe, A.M., Mining Engineer. 1,100 pages, and 902 wood-cut illustrations. 8vo, cloth........................ 10 00

WEYRAUCH (J. J.) Strength and Calculations of Dimensions of Iron and Steel Construction, with reference to the Latest Experiments. 12mo, cloth, plates............. 1 00

WILLIAMSON (R. S) On the use of the Barometer on Surveys and Reconnoissances. Part I. Meteorology in its Connection with Hypsometry. Part II. Barometric Hypsometry. With Illustrative Tables and Engravings. 4to, cloth 15 00

—— Practical Tables in Meteorology and Hypsometry, in connection with the use of the Barometer. 4to, cloth......... 2 50

---

www.ingramcontent.com/pod-product-compliance
Lightning Source LLC
Chambersburg PA
CBHW030308170426
43202CB00009B/923